普通高等教育"十四五"规划教材

材料与化工专业英语

Materials and Chemical Engineering Professional English

刘 瑶 编

北 京
冶金工业出版社
2025

内 容 提 要

本书紧密结合化学专业的核心内容和英语学习的实际需求,采用跨学科的综合设计,内容涵盖化学专业的基础知识,如元素周期表、有机化合物和化学反应等。同时,还特别介绍了实验方法和研究论文写作指南,以帮助学生在实践中学习和应用这些知识。此外,本书结合了多种教学方法,包括案例分析、实验操作指导和专业术语解析,旨在提高学生的学习兴趣和实际操作能力。

本书适合化学及相关专业的本科生和研究生阅读,也可供专业化学教师及从事材料与化工研究的专业技术人员参考。

图书在版编目(CIP)数据

材料与化工专业英语/刘瑶编. -- 北京:冶金工业出版社,2025.1. --(普通高等教育"十四五"规划教材). -- ISBN 978-7-5240-0053-2

Ⅰ. TB3;TQ

中国国家版本馆 CIP 数据核字第 2024E150A1 号

材料与化工专业英语

出版发行	冶金工业出版社	电　　话	(010)64027926
地　　址	北京市东城区嵩祝院北巷39号	邮　　编	100009
网　　址	www.mip1953.com	电子信箱	service@mip1953.com

责任编辑　王梦梦　美术编辑　吕欣童　版式设计　郑小利
责任校对　李欣雨　责任印制　禹　蕊

三河市双峰印刷装订有限公司印刷
2025年1月第1版,2025年1月第1次印刷
710mm×1000mm 1/16;7.75 印张;133 千字;116 页
定价 39.00 元

投稿电话　(010)64027932　投稿信箱　tougao@cnmip.com.cn
营销中心电话　(010)64044283
冶金工业出版社天猫旗舰店　yjgycbs.tmall.com
(本书如有印装质量问题,本社营销中心负责退换)

前　言

随着经济全球化发展，高等教育领域对培养学生专业英语的需求与日俱增。专业英语教材不仅是学生掌握专业知识的工具，更是其将来在国际舞台上进行学术交流和职业发展的重要基石。因此，编写一本结合专业知识与英语教学、适应现代教育需求的教材，已成为教育工作者和政策制定者的一项重要任务。国家主管部门和广大院校对此类教材的重视表现在各级教育政策的支持和推广上，以及在师资培训和教学资源投入上的持续增加，力图通过这些教材提升学生的专业英语能力，进而增强他们的国际竞争力。

本书旨在为材料与化工专业的学生提供一个全面的学术英语学习平台，帮助他们在学术研究、国际交流和职业发展中更有效地使用英语。本书通过紧密结合化学专业的核心内容和英语学习的实际需求，力求学生能够在理解专业知识的同时，提高英语应用能力。通过精心设计的章节内容和教学活动，帮助学生建立扎实的化学知识基础并提高英语应用能力。

本书共4章，每个章节都设计有细致的子章节，确保内容的系统性和逻辑性。第1章介绍了原子结构、元素周期表、同位素等基础知识，这些都是化学学习的基石，对于理解物质的本质和化学反应机制至关重要。第2章深入探讨了分子间作用力、不同类型的化学键合，以及多原子离子和同素异形体等，这些内容对于理解化学物质的性质和反应具有实际应用价值。第3章涵盖了实验室常见设备、安全操作和科研方法论等，旨在培养学生的实验操作能力和科研素养，这对于未来的学术研究或职业生涯极为重要。第4章系统讲解了化学反应的类型、伴随反应的现象，以及电解和金属的腐蚀等，这些都是化学工业和材料科学中的关键内容。每个章节不仅详细阐述了理论知识，还结合了当前科技发展的最新成果，如在讨论元素时引入了其在能源和材料科学中的应用，例如锂在电池技术中的重要性，或是碳纳米管和石墨烯的研究进展等。本书适合化学及相关专业的本科生和研究生使用，也可以为化学教师或从事相关研究的专业人员提供参考。

总而言之，本书是一个全面而深入的学习工具，旨在帮助化学及相关专业的学生在经济全球化的背景下，提高他们的专业英语水平，加深对化学领域的理解，并培养他们的国际视野。编者期待本书能成为学生和教师之间的桥梁，激发学生的学习热情，引领他们走向更广阔的学术与职业道路。

由于时间所限，书中不妥之处难免，欢迎读者批评指正。

编 者

2024 年 6 月

Contents

Chapter 1 Chemistry Elements ··· 1

  1.1 **Atomic structure** ··· 1

  1.2 **Periodic table** ··· 4

  1.3 **Isotopes** ··· 8

  1.4 **Alkaline metal** ·· 11

  1.5 **Alkaline earth metal** ··· 13

  1.6 **Halogen** ·· 16

  1.7 **Noble gas** ··· 19

  1.8 **Group 14 elements: C, Si, Ge, Sn, Pb** ·· 21

Chapter 2 Compounds ··· 25

  2.1 **Intermolecular forces** ··· 25

  2.2 **Bonding between atoms** ·· 28

      2.2.1 Ionic bonding ·· 28

      2.2.2 Covalent bonding ·· 32

      2.2.3 Metallic bonding ··· 34

  2.3 **Polyatomic ions** ·· 36

  2.4 **Allotropes** ··· 38

  2.5 **Soluble and insoluble** ··· 41

      2.5.1 Solution ··· 41

      2.5.2 Solubility ··· 43

      2.5.3 Insoluble ·· 44

  2.6 **Common organic compounds** ·· 48

Chapter 3   Experimental Methods ········· 55

3.1   Glassware ········· 55
3.2   General and inspection equipment ········· 58
3.3   Lab safety and maintenance ········· 61
  3.3.1   Lab safety ········· 61
  3.3.2   Lab maintenance ········· 65
3.4   How to write a research paper? ········· 68
  3.4.1   What is an expository essay? ········· 71
  3.4.2   What is a descriptive essay? ········· 74
  3.4.3   What is a narrative essay? ········· 76
  3.4.4   What is an argumentative essay? ········· 78
3.5   Research methodology ········· 81
3.6   Research integrity and ethics ········· 87

Chapter 4   Chemistry Reactions ········· 92

4.1   Displacement reactions ········· 92
4.2   Redox reaction ········· 95
4.3   Types of chemical reactions ········· 97
4.4   Electrolysis (battery) ········· 101
4.5   Corrosion of iron ········· 103
4.6   Phenomena associated with reactions ········· 105
  4.6.1   Exothermic and endothermic reactions ········· 105
  4.6.2   Crystallization ········· 108
  4.6.3   Sublimation ········· 110
  4.6.4   Catalysis ········· 112

Reference ········· 116

Chapter 1 Chemistry Elements

1.1 Atomic structure

Atoms are recognized as the foundational elements of chemistry, similar to how various ingredients compose baked goods; likewise, all matter is comprised of different types of atoms. To date, scientists have identified 118 varieties of atoms, referred to as elements, systematically arranged in the periodic table. Despite their omnipresence, individual atoms are invisibly small and cannot be detected with the naked eye.

To understand the historical belief in atoms, one must revisit ancient Greece and Democritus, credited by many historians as the first to propose the concept of indivisible units of matter. In his era, it was assumed that matter could be infinitely divided. Democritus, however, posited that there exists a finite point where matter reaches indivisible, indestructible particles, which he named "atomos" or uncuttable. Lacking empirical evidence, his theory was initially dismissed, as unsubstantiated claims were easily rejected.

Fast forwarding several centuries, the advancement in scientific extraction techniques in the Arabic world, particularly by Jabir Ibn Hayyan, evolved through meticulous experimentation that led to sophisticated methods of filtration, boiling, vapor collection, and cooling.

These experiments demonstrated that crude starting materials could be refined into exceptionally pure substances, displaying consistent properties throughout, unlike the varied mixtures found in nature. In the 1700s, French scientists Marie-Anne Paulze and Antoine Lavoisier further explored these concepts, discovering that certain pure substances could not be decomposed further through chemical reactions, affirming the notion of elements—fundamental substances that are the building blocks of all matter.

John Dalton, an early 19th-century English school teacher and chemist, observed a recurring pattern in the ratios of elements combining to form compounds, suggesting the existence of atoms—tiny indivisible units. Despite uncertainties regarding the actual size of atoms, Dalton's findings indicated uniformity in the size of atoms within an element, differing from those of other elements. His extensive work culminated in a detailed publication in 1808, which, though not immediately accepted, proved extremely useful in advancing chemical knowledge and practices.

Albert Einstein, in 1905, proposed an experimental approach and a mathematical equation to definitively confirm the existence and estimate the size of atoms. Jean Perrin, a French physicist, later validated this theory through experiments that conclusively demonstrated the reality of atoms.

Despite these advancements, visual confirmation of atoms remained elusive until the development of the Scanning Tunneling Microscope (STM) in the 1970s by engineers Gerd Bennig and Heinrich Rohrer. This technology, which employs electron tunneling to image the surface of materials at the atomic level, provided the first unambiguous images of individual atoms, subsequently refined by researchers like Dr. Wilson Ho.

Further innovations by Dr. Ara Apkarian and his team involved modifying STM technology to use light in observing atoms, a method previously thought impossible due to the incompatible scale of light wavelengths and atomic sizes. By manipulating light at the microscopic level, they successfully visualized individual atoms, thus providing direct visual evidence supporting atomic theory.

In summary, atoms are the essential units from which all chemical elements are formed, containing subatomic particles such as protons, neutrons, and electrons within a defined nucleus surrounded by an electron cloud. These elements follow a structured order in which electrons are arranged in shells around the nucleus, dictating the chemical properties and stability of the atom. The exploration and understanding of atoms have evolved significantly from ancient philosophical propositions to modern scientific confirmations, demonstrating their fundamental role in the composition of matter and the ongoing quest for scientific advancement.

New words and expressions

atoms　原子
elements　元素
periodic table　周期表
subatomic particles　亚原子粒子
nucleus　原子核
protons　质子
neutrons　中子

electrons　电子
electron cloud　电子云
chemical reactions　化学反应
compounds　化合物
scanning tunneling microscope（STM）
　　扫描隧道显微镜
electron tunneling　电子隧道效应

译文

原子被认为是化学的基础元素，正如不同的配料组成烘焙食品一样，所有物质都由不同类型的原子组成。到目前为止，科学家已经鉴定出118种原子，这些原子被称为元素，并在元素周期表中系统排列。尽管它们无处不在，但单个原子非常微小，肉眼无法看见。

为了理解历史上对原子的信仰，必须回到古希腊时期，认识德谟克利特，他是许多历史学家认为首次提出不可分割物质单位概念的人。在他所处的时代，人们认为物质可以无限分割。然而，德谟克利特提出存在一个极限，物质到达这一点时无法再被分割，他将这些粒子命名为"Atomos"或不可切割。由于缺乏实证数据，他的理论最初被驳回，因为未经证实的主张同样容易被拒绝。

数百年后，科学提取技术在阿拉伯世界取得了进步，特别是通过杰比尔·伊本·海扬的精心实验，演变成复杂的过滤、沸腾、蒸汽收集和冷却方法。

这些实验表明，粗糙的起始材料可以被精炼成极为纯净的物质，这些物质整体上显示出一致的特性，不同于自然界中常见的复杂混合物。18世纪，法国科学家夫妇玛丽·安娜·保罗兹和安托万·拉瓦锡进一步研究这些概念，发现某些纯净物质通过化学反应可以进一步分解，确认了元素的概念——构成所有物质的基本物质。

19世纪早期，英国学校教师兼化学家约翰·道尔顿观察到元素组合形成化合物的比例具有重复模式，这表明存在原子——微小的不可分割单元。尽管对原子的实际大小存在不确定性，道尔顿的发现表明，同一元素中的原子在大小上具有一致性，与其他元素的原子不同。他的广泛工作最终在1808年发表了一篇详细的出版物，尽管这些发现最初并未被普遍接受，但在推进化学知识和实践方面证明极为有用。

1905年，阿尔伯特·爱因斯坦提出了一种实验方法和数学方程，不仅证实了原子的存在，还可以精确估算原子的大小。后来，法国物理学家让·佩兰通过实验验证了这一理论，确凿地证明了原子的实际存在。

尽管取得了这些进展，但直到20世纪70年代由格尔德·贝尼格和海因里希·罗雷尔领导的工程师团队开发出扫描隧道显微镜（STM）后，人们才能够对个别原子进行无歧义的成像。这项技术利用电子隧道效应在原子级别上的成像材料表面，提供了第一幅清晰的单个原子图像，随后由研究者如威尔逊·何博士进一步完善。

更多的创新来自阿拉·阿普卡里安博士及其团队，他们改良了STM技术，使用光线观察原子，这种方法之前被认为是不可能的，因为光波长与原子大小不匹配。通过在显微镜级别操控光线，他们成功地直观显示了单个原子，从而提供了支持原子理论的直接视觉证据。

总结来说，原子是所有化学元素形成的基本单位，包含在定义好的原子核和围绕原子核的电子云中的亚原子粒子，如质子、中子和电子。这些元素遵循一种结构化的顺序，其中电子围绕核排列在壳层中，决定了原子的化学属性和稳定性。从古代哲学命题到现代科学验证，对原子的探索和理解显著进步，显示了它们在物质组成和持续的科学探索中的基本作用。

1.2　Periodic table

In the periodic table, each element is denoted by a chemical symbol, such as $^{12}_{6}C$, where "C" represents carbon. The superscript "12" indicates the mass number, and the subscript "6" denotes the atomic number. Initially, the arrangement of the periodic table may appear random, but it actually reveals intricate patterns that enhance our understanding of natural laws.

During the mid-19th century, chemists sought methods to organize elements in tabular form, proposing various formats. Dmitri Mendeleev's table was adopted due to its effective correlation of data and predictive capabilities. Mendeleev arranged elements into rows (periods) and columns (groups), grouping similar elements together. This organization not only correlated existing data but also predicted the properties of then-undiscovered elements. Mendeleev's predictions were later validated when these elements were found with properties matching his forecasts.

Currently, metals, metalloids, and nonmetals are methodically categorized in the table. It is understood that elements within the same group exhibit similar behaviors due to having identical numbers of valence electrons. For example, elements in group 1 all have one electron in their outermost shell. As one moves down the table and the principal quantum number increases, each successive element gains an additional shell, yet maintains only one electron in its outermost shell.

Several periodic trends emerge upon examination of the table. Atomic radius, for instance, increases as one moves down the table due to the addition of electron shells, but decreases across a period as nuclear charge increases, enhancing the electromagnetic attraction and reducing the atomic radius. Consequently, atomic radius generally enlarges downward and to the left of the table.

Ionic radius behaves slightly differently; adding an electron increases an ion's size, whereas removing one decreases it. Ions with identical electron configurations will have decreasing radii as atomic number increases. Ionization energy, the energy required to remove an electron, primarily from the outermost shell, decreases with the increasing distance of the electron from the nucleus, making it easier to remove. This trend is inversely related to that of atomic radius, where ionization energy increases up and to the right of the table. Francium, for instance, with its large atomic size and a single valence electron, exhibits low ionization energy due to the ease of removing the distant electron. Conversely, helium, having only one electron shell, demonstrates high ionization energy due to its compact and fully occupied shell. Successive ionization energies increase with each electron removed, especially after removing the last electron in a shell, leading to a significant increase once reaching the stable noble gas electron configuration.

Exceptions to the ionization energy trend can be explained by orbital symmetry; for example, the drop in oxygen's ionization energy compared to nitrogen is due to oxygen gaining stability upon losing an electron, contrasting with nitrogen's loss of half-filled orbital stability.

Electron affinity, the tendency of an atom to attract additional electrons, generally increases up and to the right of the table, excluding the noble gases with their filled shells. Fluorine, with the highest electron affinity, can achieve a full shell by gaining an electron. Conversely, elements in the opposite corner of the table prefer to lose electrons rather than gain them.

Lastly, electronegativity, the measure of an atom's ability to attract electrons within a chemical bond, increases up and to the right of the table. Smaller atoms like fluorine, with a higher effective nuclear charge, exhibit greater electronegativity. This property plays a crucial role in understanding chemical bonding, which will be further explored in subsequent discussions.

In summary, the periodic table not only organizes elements systematically but also elucidates their underlying atomic properties and periodic trends: atomic radius increases down and left, while ionization energy, electron affinity, and electronegativity ascend up and to the right.

New words and expressions

chemical symbol 化学符号
mass number 质量数
atomic number 原子序数
periods 周期
groups 族
valence electrons 价电子
atomic radius 原子半径
ionic radius 离子半径

ionization energy 电离能
electromagnetic attraction 电磁吸引力
electron shells 电子壳
noble gas electron configuration 稀有气体电子构型
electron affinity 电子亲和力
electronegativity 电负性
chemical bonding 化学键

译文

在周期表中，每种元素都用化学符号表示，例如$^{12}_{6}C$，其中"C"代表碳。上标"12"表示质量数，下标"6"表示原子序数。最初，周期表的排列看起来可能是随机的，但它实际上揭示了复杂的模式，增进了我们对自然规律的理解。

19世纪中期，化学家们寻求以表格形式组织元素的方法，提出了各种格式。

德米特里·门捷列夫的表格因其有效的数据相关性和预测能力而被采纳。门捷列夫将元素排列成行（周期）和列（族），将行为相似的元素分组在一起。这种组织不仅关联了现有数据，还预测了当时尚未被发现的元素的性质。门捷列夫的预测后来得到了验证，这些元素被发现具有与预期相符的性质。

如今，金属、类金属和非金属在表中被有序地分类。众所周知，同一族中元素表现出相似行为，是因为它们拥有相同数量的价电子。例如，第一族的元素都有一个电子在最外层壳。当向下移动时，主量子数增加，每个后续元素都增加了一个额外的电子壳，但最外层壳中只有一个电子。

在检视表格时，会显现几个周期性趋势。原子半径是其中之一，向下移动时原子半径增加，因为增加了电子壳。而向右移动时，由于核电荷增加，增强了电磁吸引力，原子半径减小。因此，原子半径通常向下和向左增大。

离子半径的行为略有不同；增加一个电子会使离子变大，去除一个则使其变小。具有相同电子配置的离子，其半径会随着原子序数的增加而减小。电离能是从原子中移除一个电子所需的能量，通常是从最外层壳中移除。电子离核越远，移除它就越容易。这意味着电离能的趋势与原子半径的趋势正好相反（电离能向上和向右增加）。例如，铯是一个原子较大且只有一个价电子的元素，由于电子距离核很远，因此电离容易。相反，氦只有一个电子壳，显示出高电离能，因为其壳层紧凑且完全填满。连续电离能随着移除的电子增多而增加，特别是在移除壳层中最后一个电子后，达到稳定的稀有气体电子构型后，电离能会显著增加。

电离能趋势的例外可以通过轨道对称性来解释。例如，氧的电离能相比氮有所下降，是因为氧在失去一个电子后获得了稳定性，而氮失去了半满轨道的特殊稳定性。

电子亲和力，即原子吸引额外电子的倾向，通常向上和向右增加，不考虑其壳层已满的稀有气体。氟具有最高的电子亲和力，因为如果它获得一个电子，将具有完整的壳层或稀有气体电子构型。相反，表的对角线上的元素不希望获得电子，它们更愿意失去电子。

最后，电负性是原子紧紧保持电子的能力，它将在周期表上向上和向右增加，因为像氟这样的小原子具有更多的质子及较高的有效核电荷，表现出更大的电负性。这种特性在理解化学键合方面起着至关重要的作用，这将在后续的讨论中进一步探讨。

总结来说，周期表不仅系统地组织了元素，还阐明了它们的基本原子特性和周期性趋势：原子半径向下和向左增大，而电离能、电子亲和力和电负性都向上和向右增大。

1.3　Isotopes

Understanding isotopes and their applications

Isotopes are atoms of the same element that possess identical atomic numbers, indicating an equal number of protons, but different mass numbers due to varying neutron counts. Electrons, being of negligible mass, do not affect the overall mass significantly. Thus, the variation in mass among isotopes of an element arises solely from differences in the number of neutrons. For instance, the isotopes of carbon, carbon-12 and carbon-14, share the same atomic number but have different mass numbers. Carbon-12 has a mass number of 12, including 6 protons and 6 neutrons, whereas carbon-14, with the same number of protons, includes 8 neutrons, culminating in a mass number of 14. Isotopes do not occupy separate positions on the periodic table as they share atomic numbers.

Exploring isotopes further, chlorine-35 and chlorine-37 are examples, each with 17 protons but differing in neutron counts—18 and 20 neutrons respectively, reflecting their mass numbers of 35 and 37. Another example is lithium, which has isotopes with 3 and 4 neutrons respectively, leading to mass numbers of 6 and 7 for its stable isotopic forms, Li-6 and Li-7.

Radioactive isotopes

Isotopes contribute to our understanding of nuclear stability. The larger and heavier the nucleus, the greater the need for neutrons to act as a nuclear glue, provided by the strong nuclear force, to maintain stability and prevent the protons from repelling each other. This requirement is especially pronounced in heavier elements, which need numerous neutrons to sustain nuclear cohesion.

However, some isotopes have unstable nuclei that can undergo spontaneous changes to achieve stability, a process known as radioactive decay. Such isotopes are referred to as radioactive isotopes or radioisotopes. For example, tritium, an isotope of hydrogen with two neutrons, is a radioisotope due to its unstable nucleus.

Radioisotopes have significant applications in medicine and research. They are used as tracers in medical imaging to visualize blood flow and metabolic processes, such as employing technetium-99 in diagnostic scans. Additionally, radioisotopes play a crucial role in radiotherapy for treating cancer, where they are utilized to target and destroy malignant cells.

In chemistry, isotopes serve as invaluable tools for tracking and analyzing chemical reactions. Isotopic markers enable chemists to trace reaction pathways and understand the mechanisms underlying various chemical transformations.

Summary

Isotopes of the same element share chemical properties since their chemical behavior is governed by electron configurations, unaffected by neutron variance. While some isotopes are stable, others, known as radioisotopes, undergo radioactive decay to reach a stable state. These radioisotopes are utilized extensively as tracers in medical and chemical studies, and in therapeutic applications such as cancer treatment, underscoring their importance in modern science.

New words and expressions

isotopes　同位素
atomic number　原子序数
radioactive isotopes　放射性同位素
electron configurations　电子配置
tracers　示踪剂
metabolic　代谢的
chlorine　氯

radioisotopes　放射同位素
radioactive decay　放射性衰变
nuclear glue　核黏合剂
strong nuclear force　强核力
medical imaging　医学成像
radiotherapy　放射治疗
diagnostic scans　诊断扫描

译文

了解同位素及其应用

同位素是指具有相同原子序数（即相同数量的质子）但不同质量数的同一元素的原子。由于中子数量的不同，质量数也不同。电子的质量可以忽略不计，因此不会显著影响总质量。因此，一个元素的不同同位素之间的质量差异完全由中子数量的差异造成。例如，碳的同位素碳-12 和碳-14 都具有相同的原子序数，但质量数不同。碳-12 的质量数为 12，包括 6 个质子和 6 个中子，而碳-14 虽然也有 6 个质子，却包括 8 个中子，使其质量数达到 14。同位素在周期表上不占据单独的位置，因为它们的原子序数相同。

进一步探索同位素，例如氯的两种同位素氯-35 和氯-37，每种都有 17 个质子，但中子数分别为 18 和 20，反映出它们的质量数分别为 35 和 37。另一个例子是锂，其同位素分别具有 3 个和 4 个中子，导致其稳定同位素形式的质量数分别为 6 和 7（Li-6 和 Li-7）。

放射性同位素

同位素有助于我们理解核稳定性。原子核越大越重，维持稳定性的需求就越大，需要中子充当核黏合剂，通过强核力来保持稳定，并防止质子之间的相互排斥。这种需求在重元素中特别突出，这些元素需要大量的中子来维持核的凝聚力。

然而，一些同位素具有不稳定的核，可以自发地发生变化以达到稳定状态，这个过程称为放射性衰变。这些同位素被称为放射性同位素或放射同位素。例如，氢的同位素氚就是一种放射同位素，因为其核不稳定。

放射性同位素在医学和研究中具有重要应用。它们用作医学成像的示踪剂，用于可视化血流和代谢过程，例如使用锝-99 进行诊断扫描。此外，放射同位素在放射治疗中发挥关键作用，用于治疗癌症，目的是靶向并摧毁恶性细胞。

在化学中，同位素作为跟踪和分析化学反应的宝贵工具。同位素标记使化学家能够追踪反应路径并理解各种化学变化的机制。

总结

同一元素的同位素具有相同的化学属性，因为它们的化学行为由电子配置决定，而中子的变化不会影响。虽然一些同位素是稳定的，但不稳定的同位素称为放射同位素，会经历放射性衰变以达到稳定状态。这些放射同位素广泛用作医学和化学研究中的示踪剂，也用于治疗癌症，强调了它们在现代科学中的重要性。

1.4 Alkaline metal

Group 1 of the periodic table comprises alkali metals such as lithium, sodium, potassium, rubidium, cesium, and francium. These metals are characterized by having a single electron in their valence shell, which they readily lose to form a cation with a +1 charge.

A controlled experiment with lithium, sodium, and potassium illustrates their reactivity. Due to their tendency to react with both oxygen and water vapor, forming an oxide layer, these metals are preserved under oil. Using tweezers, a piece of lithium is carefully extracted and shaped with a scalpel to expose its shiny surface and soft texture. It is essential to remove any residual oil with paper towels while avoiding skin contact. The lithium is then placed in a large trough of water, positioned behind a safety screen, to monitor its reaction. Upon immersion, lithium floats, produces bubbles, and gradually dissolves. Introducing a few drops of universal indicator solution changes the water's color to blue or purple, signifying the formation of an alkaline solution. This interaction generates hydrogen gas and lithium hydroxide, as demonstrated by the corresponding chemical equation.

Repeating this procedure with sodium reveals it to be softer and more readily cut than lithium. Sodium reacts more vigorously with water, producing hydrogen gas and sodium hydroxide. The addition of universal indicator results in a color change to blue or purple, confirming the alkaline nature of the solution. This observation highlights the reactivity of alkali metals with water, yielding an alkaline solution.

The experiment is then extended to potassium, with expectations of its softness and reactivity with water compared to sodium. It is observed that potassium is indeed softer and reacts more intensely with water. A notable aspect of the reaction is the emission of a lilac flame. Safety measures are strictly followed due to the potential risks associated with potassium's reaction, which include spitting and a distinctive "popping" sound from the combustion of hydrogen.

A distinct property of alkali metals, including potassium, is their buoyancy on water, unlike the alkaline earth metals in Group 2, which typically sink. This trend within the alkali metals group shows increasing softness and reactivity with water as one moves down the group. Rubidium and cesium are expected to demonstrate greater softness and reactivity. Francium, however, is an exception due to its radioactivity and rarity, making it an outlier in these trends. Additional observations as one progresses down the group include increasing density and decreasing melting and boiling points.

In summary, alkali metals in Group 1 react with water to produce hydrogen gas and their respective metal hydroxides. There is a marked increase in the softness and reactivity of these metals towards water down the group, accompanied by rising density and declining melting and boiling points.

New words and expressions

Group 1　第一族
alkali metals　碱金属
lithium　锂
sodium　钠
potassium　钾
rubidium　铷
cesium　铯
francium　钫
valence shell　价层
cation　阳离子
oxide layer　氧化层

universal indicator　通用指示剂
alkaline solution　碱性溶液
hydrogen gas　氢气
lithium hydroxide　氢氧化锂
sodium hydroxide　氢氧化钠
chemical equation　化学方程式
reactivity　反应性
melting point　熔点
boiling point　沸点
alkaline earth metals　碱土金属

译文

第一族的周期表包括了碱金属，如锂、钠、钾、铷、铯和钫。这些金属的特点是它们的价层中只有一个电子，很容易失去这个电子形成带有+1电荷的阳离子。

通过锂、钠和钾的实际实验，可以看到它们的反应性。由于这些金属倾向于与空气中的氧气和水蒸气反应，形成氧化层，因此它们被储存在油下。使用镊子，小心地取出一块锂并用手术刀修整，以露出其光亮的表面和柔软的质地。重要的是要用纸巾去除任何残留的油，并避免与皮肤直接接触。随后，将锂金属片放入大水槽中，并在安全屏障后观察其反应。锂与水接触后，会漂浮并产生气泡，并逐渐溶解。加入几滴通用指示剂后，水的颜色变为蓝色或紫色，表明形成了碱性溶液。这一反应产生了氢气和氢氧化锂，如化学方程式所示。

重复此过程使用钠，观察到钠比锂更软，更容易切割。钠与水的反应更为剧烈，产生氢气和氢氧化钠。再次加入通用指示剂，结果水的颜色变为蓝色或紫色，证实了溶液的碱性。这一实验观察强调了碱金属与水反应的典型性质，产生碱性溶液。

进行钾的实验时，预计其与水的软度和反应性相比钠会更显著。观察结果证实，钾确实比钠更软，与水的反应更为激烈。一个重要的观察是反应过程中发出了淡紫色的火焰。由于氢气燃烧，钾反应可能产生飞溅和可听见的"爆裂"声，因此安全措施要求使用最小的钾片，并保持足够的距离以减少风险。

碱金属的一个显著特性（包括钾在内）是它们在水上的浮力，与第二族的碱土金属不同，后者在与水反应时倾向于沉没。这揭示了碱金属组内的趋势；随着向组内下方移动，金属的软度和对水的反应性增加。预计铷和铯会显示出更大的软度和反应性。此外，由于其放射性和在自然界中的稀有性，钫在这组中是一个例外。随着向组内下方移动，还观察到密度增加及熔点和沸点降低的趋势。

总结来说，第一族的碱金属与水反应，产生氢气和相应的金属氢氧化物。在该族中，这些金属对水的软度和反应性明显增加，密度上升，熔点和沸点下降。

1.5 Alkaline earth metal

Within Group 2 of the periodic table, a trend of increasing softness among metals is observed as one descends the group. Accordingly, calcium is expected to be softer than magnesium, aligning with the general pattern where metals exhibit increased softness moving down a group. Although harder than alkali metals, calcium is notably softer than magnesium and features a shiny, silver appearance.

To experimentally examine calcium, the use of tweezers to extract a sample, followed by careful sectioning with a scalpel, proves effective. Residual oil from storage can be wiped away with paper towels. The experimental procedure involves heating a small piece of calcium with a Bunsen burner, using tongs to hold the sample in the flame. This setup allows for the observation of calcium burning with a distinctive red flame, indicating its reaction with oxygen to produce calcium oxide, a white solid. Like magnesium, calcium reacts with oxygen to form a metal oxide, underscoring their reactive qualities and confirming the predictive trends of their properties within the periodic table.

After allowing the calcium oxide to cool, a necessary safety measure due to the exothermic nature of its formation, the oxide is introduced to cold water. This interaction leads to the production of calcium hydroxide, known as limewater. The formation of a white precipitate, slightly soluble in water, settling at the bottom of the beaker, is observed.

In contrast, magnesium reacts only with steam, not cold water. This reaction produces magnesium oxide and hydrogen gas. Essential to this experiment is the direct heating of the magnesium strip, while avoiding the water-soaked mineral wool. Hydrogen gas generated during this process can be collected over water for subsequent verification.

Comparing the reactivity of calcium with water to that of magnesium reveals a more vigorous reaction from calcium. This is in line with the observed trend of increased reactivity among metals further down Group 2. The reaction of calcium with cold water produces calcium hydroxide and releases hydrogen gas, with visible effervescence indicating gas production and the metal characteristically sinking to the bottom of the reaction vessel.

In conclusion, the alkaline earth metals of Group 2, including magnesium and calcium, demonstrate reactivity trends similar to alkali metals. Both react with oxygen to form their respective oxides. Magnesium reacts with steam to yield magnesium oxide and hydrogen gas, while calcium directly engages with cold water, resulting in the formation of calcium hydroxide and hydrogen gas.

New words and expressions

Group 2　第二族
tongs　钳子
metals　金属
tweezers　镊子
scalpel　手术刀
bunsen burner　本生灯
calcium oxide　氧化钙
metal oxides　金属氧化物

exothermic reaction　放热反应
calcium hydroxide　氢氧化钙
limewater　石灰水
precipitate　沉淀
steam　蒸汽
magnesium oxide　氧化镁
mineral wool　矿物棉
effervescence　气泡

译文

在周期表的第二族中，观察到一种趋势，即随着组内向下移动，金属的柔软性逐渐增加。因此，预期钙比镁更柔软，这与金属向下移动时柔软性增加的一般模式一致。虽然钙和镁比碱金属硬，但钙确实比镁更柔软，并且具有闪亮的银色外观。

为了实验性地检验钙，使用镊子提取样本，然后小心地用手术刀切割是一种有效的方法。可以用纸巾擦去储存时残留的油。实验程序包括使用镊子将小块钙放入本生灯的火焰中加热。此设置允许观察到钙燃烧时发出独特的红色火焰，表明它与氧气反应生成白色固体氧化钙。与镁一样，钙与氧气反应形成金属氧化物，凸显它们的反应性质，并证实了周期表内它们性质的预测趋势。

在氧化钙冷却后（由于其形成的反应是放热的，冷却是一个必要的安全措施），将氧化物置于冷水中。这一作用导致氢氧化钙（通称石灰水）的形成。观察到在试管底部形成的白色沉淀物，这些沉淀物在水中略微可溶。

与此形成对比的是，镁只与蒸汽反应，不与冷水反应。这一反应产生氧化镁和氢气。进行这一实验时，直接加热镁条是关键，同时避免接触浸水的矿物棉。在此过程中产生的氢气可以收集于水上，以进一步测试以确认其存在。

比较钙与水的反应性与镁的反应性，预计钙会表现出更剧烈的反应。这与组内金属向下移动时反应性增加的趋势相符。钙与冷水的反应不仅生成氢氧化钙，还释放氢气，反应器底部沉淀的钙特征性地显示出产气的气泡。

总结来说，第二族的碱土金属包括镁和钙，展示了与碱金属相似的性质和反应性趋势。镁与蒸汽反应生成氧化镁和氢气，而钙则直接与冷水反应，生成氢氧化钙和氢气。

1.6 Halogen

Halogens, located in Group 7 of the periodic table, include fluorine, chlorine, bromine, iodine, and astatine. This discussion will cover the key reactions of halogens: displacement reactions, reactions with metals, and reactions with hydrogen.

Halogens in their elemental state are toxic, yet their compounds have numerous practical applications. All halogens possess seven electrons in their valence shell and can gain an electron to form an anion with a minus one charge. As one moves down the group, atomic radii expand due to the addition of electron shells, which correspondingly results in increases in melting points, boiling points, and densities. For instance, fluorine and chlorine exist as gases, bromine as a liquid, and iodine and astatine as solids. Fluorine appears as a very pale green gas, chlorine as a greenish-yellow gas, bromine as a reddish-brown liquid, iodine as a gray solid that sublimates to a purple vapor, and astatine as a black solid that is radioactive.

Reactivity decreases down the group, meaning a halogen higher in the group will always be more reactive than those below it. This reactivity allows a more reactive halogen to displace a less reactive one from a compound. To illustrate, consider displacement reactions with aqueous solutions of potassium chloride, potassium bromide, and potassium iodide. Adding chlorine water to potassium chloride elicits no reaction. However, chlorine, being more reactive than bromine, displaces bromine in potassium bromide, resulting in the formation of potassium chloride and free bromine. The solution turns reddish-brown due to the released bromine. Similarly, adding chlorine water to potassium iodide results in a displacement reaction that frees iodine, producing a dark brown solution due to the iodine's presence.

Examining what occurs when bromine water is added to the same three solutions reveals that adding bromine water to potassium chloride causes the solution to turn reddish-brown, but no displacement reaction occurs since bromine is less reactive than chlorine. No reaction is noted when bromine water is added to potassium bromide. However, since bromine is more reactive than iodine, it displaces iodine in potassium iodide, resulting in a dark brown solution. If iodine were added to these solutions, no displacement reactions would occur, as iodine is less reactive than both chlorine and bromine.

Halogens also react with certain metals to form metal halides; for example, sodium chloride, commonly known as table salt, is a prevalent metal halide extracted from the sea or salt mines. Sodium chloride results from the reaction of heated sodium metal with chlorine gas. This reaction is highly exothermic, and thus, not recommended for laboratory settings.

Additionally, halogens react with hydrogen gas to produce hydrogen halides, most of which are highly exothermic reactions. For example, chlorine gas and hydrogen gas can combine to form hydrogen chloride.

In summary, the Group 7 halogens participate in displacement reactions where a more reactive halogen displaces a less reactive one from a compound. They also react with metals to form metal halides and with hydrogen to yield hydrogen halides.

New words and expressions

halogens   卤素
Group 7   第七族
fluorine   氟
bromine   溴
iodine   碘
astatine   砹
displacement reactions   置换反应
hydrogen   氢
electron shell   电子层
atomic radii   原子半径

density   密度
sublimates   升华
reactive   反应性
anion   阴离子
metal halides   金属卤化物
sodium chloride   氯化钠
exothermic reaction   放热反应
hydrogen halides   氢卤化物
hydrogen chloride   氯化氢

译文

卤素位于周期表的第七族，包括氟、氯、溴、碘和砹。本节讨论卤素的主要反应：置换反应、与金属的反应及与氢的反应。

卤族元素状态下是有毒的，但其化合物在实际应用中却非常广泛。所有卤素的价层中都含有 7 个电子，并且能够获得 1 个电子形成带负电荷的阴离子。随着在族中向下移动，由于电子层的增加，原子半径、熔点、沸点及密度也随之增大。例如，氟和氯是气体，溴是液体，碘和砹是固体。氟呈非常浅的绿色气体，氯呈黄绿色气体，溴是红棕色液体，碘是灰色固体并能升华成紫色蒸气，砹是黑色固体且具有放射性。

随着族内向下移动，卤素的反应性逐渐降低，因此位于更高位置的卤素总是比下面的卤素更具反应性。一个更具反应性的卤素将总是能够在化合物中置换掉一个较不具反应性的卤素。例如，当将氯水加入氯化钾、溴化钾和碘化钾的水溶液中时，加入氯化钾的氯水不会发生反应。但由于氯比溴更具反应性，氯将在溴化钾中置换溴，生成氯化钾并释放溴，溶液因此呈现红棕色。类似地，当氯水加入碘化钾溶液中时，由于氯比碘更活泼，将发生置换反应，释放出碘，溶液因碘的存在而呈现深棕色。

考虑加入溴水时，当溴水加入氯化钾溶液时，会使溶液变为红棕色，但由于溴的反应性不如氯，不会发生置换反应。当溴水加入溴化钾溶液时也不会发生反应。由于溴比碘更具反应性，它将在碘化钾中置换碘。因此，由于碘的存在，生成了深棕色溶液。如果将碘加入这三种溶液中，则不会发生置换反应，因为碘的反应性不如氯和溴。

卤素还可以与特定金属反应生成金属卤化物；例如，氯化钠（普通的食盐）就是一种常见的金属卤化物，从海洋或盐矿中提取。氯化钠可以通过加热钠金属与氯气反应来制得，由于这是一个高度放热的反应，在实验室中通常不推荐进行此反应。

此外，卤素还可以与氢气反应生成相应的氢卤化物，这些反应大多也是高度放热的。例如，氯气和氢气可以反应生成氯化氢。

总结来说，第七族的卤素参与置换反应，其中更具反应性的卤素会在化合物中置换较不活泼的卤素。卤素还可以与金属反应生成相应的金属卤化物，以及与氢反应生成氢卤化物。

1.7　Noble gas

Noble gases, which comprise Group 18 of the periodic table, include helium, neon, argon, krypton, xenon, and radon. These elements are distinguished by their complete valence electron shells; helium has two valence electrons, while the others possess eight. This electron configuration imparts significant stability to the noble gases, historically regarded as entirely unreactive and thus often labeled as inert gases. However, advancements in chemistry have shown that under certain conditions, some noble gases can form compounds, indicating a more complex reactivity than previously understood.

Noble gases are colorless and exist in a monatomic state, meaning they are comprised of single atoms. An increase in electron shells is observed as one moves down the group, which leads to larger atomic sizes. This growth in size results in higher boiling points due to strengthened intermolecular forces among larger atoms with more electrons. Furthermore, the density of these gases also increases down the group as the larger atoms occupy more volume.

Despite their renowned inertness, noble gases are extensively utilized in various practical applications. Helium, being lighter than air, is used to inflate party balloons and airships due to its buoyant properties. Krypton and xenon find uses in advanced laser technologies and the manufacturing of flat-panel displays. Particularly from the 1990s onward, xenon has been increasingly used in automotive headlights to improve road safety through enhanced illumination. These applications underscore the indispensable roles noble gases play in technology and everyday life.

Xenon lamps, known for emitting intensely bright light, significantly improve contrasts and color perception. Neon, when activated by an electrical current, emits a vivid orange glow, making it ideal for lighting applications. Argon is commonly used in light bulbs to prevent the tungsten filament from reacting, benefiting from its greater inertness compared to air.

In the medical field, xenon is valued as an anesthetic that stabilizes blood pressure and heart rate with minimal side effects during surgeries. Radon, notable for its high radioactivity, is used in radiotherapy for cancer treatment, exploiting its radioactive decay for therapeutic purposes. Additionally, krypton is utilized in double glazing between glass panels to enhance insulation, taking advantage of its lower thermal conductivity compared to air.

In the aerospace industry, xenon serves as a propellant for satellites, its substantial mass crucial for adjusting satellite trajectories along orbital paths. These examples illustrate the broad utility and significance of the noble gases—also known as Group zero—highlighting their stability and versatility across a spectrum of applications from everyday technologies to specialized uses in medicine and space exploration.

New words and expressions

noble gases　卤素
Group 18　第十八族
helium　氦
neon　氖
argon　氩
krypton　氪
xenon　氙
radon　氡
valence electrons　价电子
monatomic　单原子
intermolecular forces　分子间力
density　密度
laser technologies　激光技术
flat-panel displays　平板显示器
automotive headlights　汽车前灯
xenon lamps　氙气灯
electrical current　电流
light bulbs　灯泡
tungsten filament　钨丝灯丝
anesthetic　麻醉剂
radiotherapy　放射治疗
radioactive decay　放射性衰变
double glazing　双层玻璃
thermal conductivity　热导率
propellant　推进剂
satellites　卫星
orbital paths　轨道路径

译文

稀有气体属于元素周期表的第18族，包括氦、氖、氩、氪、氙和氡。这些元素的特点是它们的价电子壳层已满；氦有两个价电子，而其他元素有8个。这种电子配置赋予了稀有气体显著的稳定性，他们曾被认为完全不活跃，因此常被称为惰性气体。然而，化学的进步显示，在特定条件下，一些稀有气体可以形成化合物，表明它们的反应性比以前理解得更为复杂。

稀有气体是无色的，并且以单原子状态存在，意味着它们由单个原子组成。在族内向下移动时观察到电子壳的增加，这导致原子尺寸变大。这种尺寸的增长导致沸点升高，因为较大的原子拥有更多电子，从而增强了分子间的力。此外，这些气体的密度也随着较大的原子占据更多体积而增加。

尽管稀有气体以其惰性而著称，它们在各种实际应用中被广泛使用。由于氦比空气轻，它被用来充气派对气球和飞艇。氖和氙在先进的激光技术和平板显示器制造中找到了用途。特别是从20世纪90年代起，氙在汽车前灯中的使用越来越广泛，它通过提高照明效果来提高道路安全。这些应用凸显了稀有气体在技术和日常生活中不可或缺的角色。

氖灯以其强烈的亮光而闻名，可显著改善对比度和颜色感知。当电流激活氖时，它发出鲜明的橙色光芒，非常适合照明应用。氩通常用于灯泡中，利用其比空气更大的惰性，防止钨丝反应。

在医疗领域，氙被用作一种麻醉剂，可在手术期间稳定血压和心率，且副作用最小。因高放射性而著名的氡在癌症治疗中的放疗中被使用，利用其放射性衰变进行治疗。此外，氪在双层玻璃中用于提高绝热性，是利用其比空气低的热导率。

在航天工业中，氙用作卫星的推进剂，其可观的质量对于调整卫星沿轨道路径的轨迹至关重要。这些例子展示了稀有气体的广泛用途和重要性，也称为第零族，突出显示了它们在从日常技术到医学和空间探索的专业用途的稳定性和多功能性。

1.8　Group 14 elements：C，Si，Ge，Sn，Pb

Each element within Group 14 of the periodic table has four electrons in the outermost shell of its atoms, adhering to the electronic configuration $ns^2\ np^2$, where "n" represents an integer greater than one. Despite these elements forming compounds with often similar structures, there is a discernible evolution in properties as one progresses from carbon to lead.

Carbon stands out prominently among these elements, primarily existing in two distinct allotropic forms: diamond and graphite. This versatility extends into the realm of organic chemistry, which explores the vast array of compounds that carbon can form. This capability is largely attributed to carbon's exceptional ability for catenation, where atoms of the same element form long chains through covalent bonds.

Moving down the group, silicon and germanium occupy the boundary between metals and non-metals, classified as metalloids. Both elements share crystalline structures reminiscent of diamond, underscoring their semi-metallic nature. Tin, further down the group, exhibits allotropic versatility, transitioning through different physical states. In contrast, lead primarily appears in a metallic form, marking a shift from the allotropic diversity observed in its predecessors.

This progression from carbon to lead encapsulates a journey through varying physical states and chemical behaviors, from the intricate organic chemistry rooted in carbon's ability to form diverse molecular structures, to the semi-conducting properties of silicon and germanium, and finally to the metallic characteristics of tin and lead. This group exemplifies the profound impact of electronic configuration on the elements' chemical and physical properties, bridging organic chemistry and materials science.

In the domain of chemistry, catenation—the ability of atoms to form chains by bonding with themselves—is particularly notable in carbon. Unlike carbon, the propensity for catenation significantly diminishes in silicon and germanium. Although these elements can form hydrides similar to simpler hydrocarbons, their stability is comparatively less robust. The essence of catenation for silicon and germanium lies in the formation of Si—Si and Ge—Ge covalent bonds. These bonds, being inherently longer and weaker than the robust C—C bonds in carbon compounds, highlight carbon's unique ability to establish strong, stable chains.

Moreover, carbon is unique within its group for its ability to form multiple bonds not only with other carbon atoms but also with different elements like oxygen. This ability greatly expands the diversity and complexity of compounds carbon can form, a trait not mirrored to the same extent by its group counterparts.

Turning to silica (SiO_2), despite a formula that suggests a simple diatomic molecule, its structural complexity is considerably more intricate. Unlike carbon, silicon cannot form multiple bonds with oxygen, instead favoring the formation of single covalent bonds with four oxygen atoms in a tetrahedral configuration. This structure is not isolated but repeats across numerous units, forming a network solid. This complex assembly parallels the durability and interconnectedness observed in diamond, resulting in a material with a high melting point. The various forms of silica, all adhering to this fundamental tetrahedral motif, underscore the stark contrast in structural and bonding capabilities between carbon and its group counterparts, illustrating carbon's unique chemical versatility within its group.

New words and expressions

Group 14　第十四族
electronic configuration　电子配置
allotropic forms　同素异形体
organic chemistry　有机化学
catenation　成链能力
covalent bonds　共价键
metalloids　类金属
graphite　石墨
silicon　硅
crystalline structures　结晶结构
semi-conducting properties　半导体特性
hydrides　氢化物
hydrocarbons　碳氢化合物
silica（SiO_2）　二氧化硅
diatomic molecule　二原子分子
tetrahedral configuration　四面体结构
network solid　网络固体
germanium　锗
lead　铅
tin　锡

译文

　　第十四族的每个元素在其原子的最外层壳中都有 4 个电子，遵循电子配置 $ns^2 np^2$，其中 n 表示大于 1 的整数。尽管这些元素形成的化合物往往结构相似，但从碳到铅，其性质可观察到了显著的演变。

　　碳在这些元素中极为突出，主要以两种独特的同素异形体存在：金刚石和石墨。这种多样性延伸到有机化学领域，探索碳能形成的无数化合物。这主要归功于碳的卓越成链能力，即同一元素的原子通过共价键形成长链。

　　向下移动到族中的硅和锗，它们处于金属和非金属之间的边界，被归类为类金属。两者都具有类似金刚石的结晶结构，凸显它们的半金属特性。更下面的锡显示了同素异形体的多样性，通过不同的物理形态过渡。相比之下，铅则完全以金属形态出现，标志着与其族中前辈们所见同素异形体多样性的一种背离。

从碳到铅的这一过程概括了从复杂的有机化学（碳具有形成多样分子结构的能力）到硅和锗的半导体特性，再到锡和铅的金属特性的物理状态和化学行为的变化。这一族展示了电子配置对元素化学和物理性质的深远影响，架起了有机化学与材料科学之间的桥梁。

在化学领域，成链能力——原子通过与自身结合形成链的能力——在碳中尤为突出。与碳不同，硅和锗的成链倾向显著减弱。虽然这些元素确实能够形成类似简单碳氢化合物的氢化物，但它们的稳定性相对较弱。硅和锗的成链本质在于形成 Si—Si 和 Ge—Ge 的共价键。这些键本质上较长且较弱，与碳化合物中坚固的 C—C 键相比，凸显了碳建立强大、稳定链的独特能力。

此外，碳是其族中唯一能够与其他碳原子及不同元素（如氧）形成多重键的元素。这一能力显著扩展了碳能形成的化合物的多样性和复杂性，这一特性未能在同族的其他元素中同样体现。

转向二氧化硅（SiO_2），尽管其化学式暗示了一个简单的二原子分子，其结构复杂性却远超预期。与碳不同，硅无法与氧形成多重键，而是倾向于与 4 个氧原子形成单一的共价键，构成四面体结构。这种结构不是孤立的，而是在许多单元中重复出现，形成网络固体。这种复杂的组装与金刚石结构中看到的耐用性和相互连接性相呼应，产生了一个高熔点的材料。二氧化硅的多种形态都遵循这一基本的四面体模式，突出了碳与其族中其他元素在结构和键合能力上的鲜明对比，展示了碳在其元素族中所持有的独特化学多样性。

Chapter 2 Compounds

2.1 Intermolecular forces

This section elucidates the concept of intermolecular forces, outlining three prevalent types and delineating the distinctions among them. An intermolecular force is an attractive interaction occurring between adjacent molecules. The three principal types of intermolecular force are: permanent dipole-dipole forces, hydrogen bonds, and Van der Waals forces. These forces are significantly weaker than the ionic or covalent bonds that bind atoms and ions within elements and compounds.

Focusing first on permanent dipole-dipole forces, a molecule is considered polar if it exhibits a permanent dipole, typically resulting from varying electronegativities among its constituent atoms. Electronegativity refers to the affinity of an atom within a molecule for the electron pair constituting a covalent bond. Hydrogen chloride serves as an illustrative example, classified as a polar molecule due to the electron pair in the hydrogen-chlorine bond being more proximal to the chlorine atom, thereby creating a dipole. The chlorine atom, possessing a higher electronegativity compared to the hydrogen atom, exhibits a stronger pull on the electron pair. This polarity can be represented with hydrogen marked as $\delta+$ and chlorine as $\delta-$, facilitating an attraction between the $\delta-$ chlorine atom of one molecule and the $\delta+$ hydrogen atom of another.

The hydrogen bond constitutes the second category of intermolecular forces. In this type, the permanent dipole arising from a covalent bond between a hydrogen atom and an atom of fluorine, oxygen, or nitrogen is exceptionally strong. Consequently, the attraction between the electron-deficient $\delta+$ hydrogen of one molecule and the lone pair of electrons on a fluorine, oxygen, or nitrogen atom of another molecule surpasses the strength of the previously mentioned permanent dipole-dipole attraction, such as that observed between two hydrogen chloride molecules. This specific form of dipole-dipole interaction, involving an electron-deficient δ + hydrogen of one molecule and the lone pair on a fluorine, oxygen, or nitrogen atom of another molecule, is distinctively termed a hydrogen bond. Therefore, a hydrogen bond is defined as the attraction between the δ + hydrogen of one molecule and the δ − associated with the lone pair of a fluorine, oxygen, or nitrogen atom of an adjacent molecule. Although the strength of a hydrogen bond is only about 5% of that of a covalent bond, its impact on the physical properties of compounds is profound. For instance, in the absence of hydrogen bonds, substances like water and alcohol would exist as gases under standard room temperature and pressure.

The third category, Van der Waals forces, are characterized as induced dipole-dipole interactions. These arise from the movement of electrons within atomic shells, creating an instantaneous dipole at any given moment. Van der Waals forces are universal, affecting all molecules irrespective of their polarity, and stand as the sole intermolecular force present among nonpolar molecules, including the halogens and noble gases. The strength of Van der Waals forces escalates with the increase in the number of electrons in a molecule, elucidating the observed rise in boiling points within the halogen group and among noble gases.

In conclusion, intermolecular forces represent the attractive interactions between neighboring molecules, with the three principal types being permanent dipole-dipole forces, hydrogen bonds, and Van der Waals forces.

New words and expressions

intermolecular forces　分子间力
permanent dipole-dipole forces　永久偶极-偶极力
hydrogen bonds　氢键
Van der Waals forces　范德瓦尔斯力
ionic bonds　离子键
polar molecule　极性分子
electronegativity　电负性
electron pair　电子对
dipole　偶极

lone pair	孤对电子	atomic shells	原子壳层
induced dipole-dipole interactions		halogens	卤素
诱导偶极-偶极相互作用		noble gases	稀有气体

译文

  本节阐述了分子间力的概念，并概述了 3 种常见的分子间力，阐明了它们之间的区别。分子间力是相邻分子之间的吸引作用。三种主要类型的分子间力分别是：永久偶极-偶极力、氢键和范德瓦尔斯力。这些力都显著弱于形成元素和化合物中原子和离子间的离子键或共价键。

  首先关注永久偶极-偶极力，如果一个分子表现出永久偶极，通常是由于其构成原子之间的电负性不同，这样的分子被认为是极性分子。电负性指的是分子中原子对构成共价键的电子对的亲和力。以氯化氢为例，由于氢—氯键中的电子对更靠近氯原子，因此形成了一个偶极，氯原子由于比氢原子具有更高的电负性，表现出对电子对的更强拉力。这种极性可以用氢标为 $\delta+$ 和氯标为 $\delta-$ 来表示，从而促进了一个分子中的 $\delta-$ 氯原子与另一个分子中的 $\delta+$ 氢原子之间的吸引。

  氢键是第二种类型的分子间力。在这种类型中，由氢原子与氟、氧或氮原子之间的共价键形成的永久偶极特别强。因此，一个分子中电子缺失的 $\delta+$ 氢与另一个分子中的氟、氧或氮原子的孤对电子之间的吸引力超过了上述永久偶极-偶极力的强度，例如氯化氢分子之间的力。这种特定形式的偶极-偶极相互作用，涉及一个分子中电子缺失的 $\delta+$ 氢和另一个分子中氟、氧或氮原子的孤对电子，被特别称为氢键。因此，氢键被定义为一个分子中的 $\delta+$ 氢与相邻分子中带孤对电子的氟、氧或氮原子的 $\delta-$ 之间的吸引力。尽管氢键的强度只有共价键的约 5%，但它对化合物的物理性质有深远的影响。例如，如果没有氢键，像水和酒精这样的物质在标准室温和压力下会以气体形态存在。

  第三种类型，范德瓦尔斯力，被描述为诱导偶极-偶极相互作用。这种力源于原子壳层内电子的运动，任何给定时刻都会产生瞬时偶极。范德瓦尔斯力普遍影响所有分子，无论它们的极性如何，并且是非极性分子（包括卤素和稀有气体）之间唯一存在的分子间力。随着分子中电子数量的增加，范德瓦尔斯力的强度增加，这解释了卤素族和稀有气体中沸点的升高。

  总之，分子间力代表了相邻分子之间的吸引相互作用，其中 3 种主要类型是永久偶极-偶极力、氢键和范德瓦尔斯力。

2.2 Bonding between atoms

2.2.1 Ionic bonding

Within the framework of chemical bonding, Lewis symbols are employed to visually represent the electron configurations of atoms, highlighting only the valence (outermost) electrons through the use of dots and crosses. These valence electrons are pivotal, as they are exclusively involved in the formation of chemical bonds. In the illustrative reaction between a sodium atom and a chlorine atom, the sodium atom donates an electron to the chlorine atom. This electron transfer enables the sodium atom to emulate the stable electron configuration of neon, while chlorine, by acquiring an additional electron, mirrors the stable electron configuration of argon.

Prior to the reaction, the sodium atom possesses a balanced count of 11 electrons and 11 protons, rendering it electrically neutral. The donation of an electron to chlorine disrupts this balance, leaving the sodium atom with 10 electrons and 11 protons, thus imparting it with a positive charge equivalent to the charge of one proton (+1). The resultant species, denoted as Na^+, is referred to as a sodium ion. Conversely, the chlorine atom, upon accepting an electron, encompasses 17 protons and 18 electrons, leading to an overall negative charge of -1. This entity is denoted as Cl^- and termed a chloride ion. Notably, the transferred electron becomes indistinct from chlorine's other valence electrons, collectively constituting the outer electron shell of the chloride ion. The symbols of crosses and dots serve merely to trace the origins of the electrons, without suggesting any inherent differences between them.

Sodium chloride, the product of this reaction, consists of Na^+ and Cl^- ions, combined in a stoichiometric ratio of 1 : 1, defining it as an ionic compound. The nature of bonding within such compounds is ionic, a characteristic of structures formed through the transfer of electrons from one atom (or group of atoms) to another. It is essential to recognize that, although sodium and chlorine attain electron configurations resembling those of noble gases post-reaction, they do not transmute into noble gases.

The essence of an element is dictated by its proton count, which remains unaltered in both sodium and chlorine throughout the reaction, preserving their elemental identities.

Ionic compounds exhibit a set of characteristic properties that are universally observed:

(1) High melting and boiling points (non-volatile nature): Typically, ionic compounds exist as solids at ambient temperature due to their high melting and boiling points. This high thermal threshold is attributed to the substantial energy required to disrupt the ionic lattice structure. The ions within this structure are bound by strong electrostatic forces, necessitating significant energy input to overcome these attractions and transition the compound from solid to liquid or gas.

(2) Solubility in water: Water, a polar solvent, has molecules featuring a charge-separated nature, known as a dipole. This polarity enables water to effectively disrupt the crystal lattice of ionic compounds, facilitating their dissolution. The process involves water molecules engaging with the ions of the lattice, drawing them into solution and thereby disintegrating the lattice. Specifically, the positively charged region of a water molecule can attract and detach an anion from the lattice, while its negatively charged region can similarly influence a cation.

(3) Electrical conductivity in molten state or aqueous solution: To conduct electricity, a substance must contain mobile charged particles. Whereas solid ionic compounds are electrical insulators due to their fixed ions within the strong electrostatic lattice forces, these compounds become conductive when either melted or dissolved in water. In such states, the ions are free to move, allowing them to transport electrical current, thus classifying the substance as an electrolyte.

(4) Fragility of ionic crystals: Ionic crystals are prone to shattering under applied force. This vulnerability arises when force induces the displacement of ion layers within the crystal, causing ions of like charges to align adjacently. The repulsive forces between these similarly charged ions destabilize the crystal structure, leading to its fragmentation.

(5) Conditions favoring ionic bonding: Ionic bonding predominantly occurs between metals, which have relatively low ionization energies, and non-metals. Metals, often with no more than three electrons in their outer shell, readily lose electrons, while non-

metals, typically with more than three valence electrons, tend to gain electrons. However, despite having only one electron in its outer shell, hydrogen is an exception due to its high ionization energy, attributed to the electron's proximity and strong attraction to the nucleus. Consequently, hydrogen is classified as a non-metal, and most of its compounds do not exhibit ionic character.

New words and expressions

chemical bonding　化学键
Lewis symbols　路易斯符号
electron configurations　电子构型
ionic compound　离子化合物
sodium ion　钠离子
chloride ion　氯离子
stoichiometric ratio　化学计量比
electrostatic forces　静电力
lattice　晶格
insulator　绝缘材料
solubility　溶解度

fragility　脆性
polar solvent　极性溶剂
electrical conductivity　电导率
electrolyte　电解质
ionic crystals　离子晶体
ionization energies　电离能
non-metals　非金属
proximity　接近
thermal threshold　温度阈值
molten　熔融
aqueous　水溶剂的

译文

在化学键的框架中，路易斯符号用于直观地表示原子的电子构型，通过使用点和叉只突出显示最外层的价电子。这些价电子在形成化学键中起着至关重要的作用。在钠原子和氯原子之间的示例反应中，钠原子向氯原子捐赠一个电子。这种电子转移使钠原子模仿稳定的氖电子构型，而通过获取额外的电子，氯则模仿了稳定的氩电子构型。

在反应之前，钠原子拥有 11 个电子和 11 个质子，使其电荷中性。向氯捐赠一个电子打破了这种平衡，使钠原子拥有 10 个电子和 11 个质子，因此赋予了它相当于一个质子电荷的正电荷（+1）。结果生成的物种被标记为 Na^+，称为钠离子。相反，氯原子在接受一个电子后，包含 17 个质子和 18 个电子，导致总负电荷为 -1。这个实体被标记为 Cl^-，称为氯离子。值得注意的是，转移的电子与氯的其他价电子无法区分，共同构成了氯离子的外电子壳。叉和点的符号仅用来追踪电子的来源，并不暗示它们之间有任何固有的差异。

由此反应生成的氯化钠由 Na^+ 和 Cl^- 组成，以 1∶1 的化学计量比结合，定义为一种离子化合物。这类化合物的键合性质为离子性，这是通过一个原子（或原子组）向另一个原子转移电子形成的结构的特征。重要的是要认识到，尽管钠和氯在反应后获得了类似于稀有气体的电子构型，它们并没有转变为稀有气体。元素的本质由其质子数量决定，这在整个反应过程中钠和氯都保持不变，保留了它们的元素身份。

离子化合物表现出一组普遍观察到的特征性质：

（1）高熔点和沸点（非挥发性）：通常，由于它们的高熔点和沸点，离子化合物在室温下以固体存在。这种高热阈值归因于破坏离子晶格结构所需的大量能量。晶格内的离子通过强静电力相互束缚，需要显著的能量输入来克服这些吸引力，并将化合物从固体转变为液体或气体。

（2）水中的溶解度：水是一种极性溶剂，具有分离电荷的分子，称为偶极。这种极性使水能有效地破坏离子化合物的晶格结构，促进它们的溶解。这一过程涉及水分子与晶格的离子相互作用，将它们吸引到溶液中，从而瓦解晶格。具体来说，水分子的带正电区域可以吸引并分离晶格中的阴离子，而其带负电区域可以类似地影响阳离子。

（3）熔融状态或水溶液中的电导性：要传导电流，物质必须包含可移动的带电粒子。虽然固态离子化合物由于其离子在强静电晶格力中固定而是电绝缘体，但这些化合物在熔化或溶解于水时变为导体。在这些状态下，离子可以自由移动，允许它们传输电流，因此将该物质归类为电解质。

（4）离子晶体的脆性：离子晶体容易在施加力时破碎。当力导致晶体内的离子层位移，使同电荷的离子相邻排列时，这种脆性产生。这些同电荷离子之间的排斥力破坏了晶体结构，导致其破裂。

（5）有利于离子键形成的条件：离子键主要发生在电离能相对较低的金属和非金属之间。金属通常在其外壳中最多有 3 个电子，容易失去电子，而非金属通常有超过 3 个价电子，倾向于获得电子。然而，尽管氢在其外壳中只有一个电子，由于电子与核的接近性和强烈的吸引力，其电离能很高，因此氢被归类为非金属，其大多数化合物不表现出离子特性。

2.2.2 Covalent bonding

When the energy required to remove electrons from an atom involved in bonding is prohibitively high, the atoms—typically non-metals—opt to share electrons to attain a stable electronic arrangement akin to that of noble gases. This electron-sharing mechanism is termed covalent bonding. For instance, in the formation of a chlorine molecule, no ions are produced. To differentiate between the electrons from each chlorine atom, one set of electrons is marked with dots and the other with crosses. However, once these electrons partake in the covalent bond, they become indistinguishable. Consequently, chlorine gas, with the chemical formula Cl_2, manifests as a diatomic molecule, enabling the chlorine atoms to mimic the stable electronic configuration of a noble gas. The chlorine molecule is denoted as Cl—Cl, where the "—" symbolizes a shared electron pair, representing a covalent bond. An atom's or group's valency, or its number of "hooks", simplifies the description of how many electrons it needs to either donate, accept, or share to achieve the stability of a noble gas's electronic structure. Take carbon, with an electronic structure of 2.4, as an example; it's said to have a valency of four, meaning it requires sharing four electrons to reach the stable configuration of neon, which is 2.8. In certain instances, atoms share more than one electron pair. For example, each oxygen atom shares two electron pairs to attain neon's electron configuration, represented by the structural formula O=O, indicating a double bond. Similarly, a triple bond forms when three pairs of electrons are shared.

Characteristics of covalent compounds include:

(1) Volatility with low melting and boiling points: These compounds often appear as liquids or gases at room temperature (e.g., O_2, H_2O, N_2, CO_2) due to the absence of strong electrostatic attractions between molecules, allowing them to separate at lower temperatures. The weak forces binding these molecules together are known as van der Waals' forces.

(2) Limited water solubility: Without ions in their structure, covalent compounds mix less easily with water molecules. They tend, however, to dissolve well in non-polar organic solvents (like benzene) that also consist of covalent molecules.

(3) Non-conductivity of electricity: The absence of ions in covalent compounds means they cannot conduct electricity, contrasting with ionic compounds that can facilitate electrical flow.

New words and expressions

ions 离子
diatomic molecule 二原子分子
valency 化合价
electronic structure 电子结构
double bond 双键
triple bond 三键
volatility 挥发性
water solubility 水溶性
non-polar organic solvents 非极性有机溶剂

译文

当从原子中移除电子所需的能量过高时，这些原子——通常是非金属——选择共享电子以达到类似于稀有气体的稳定电子排布。这种电子共享机制被称为共价键。例如，在形成氯分子时，不产生离子。为了区分来自每个氯原子的电子，一组电子用点表示，另一组用叉表示。然而，一旦这些电子参与共价键，它们就变得无法区分。因此，氯气以化学式 Cl_2 表现为一个二原子分子，使氯原子模仿稀有气体的稳定电子构型。氯分子表示为 Cl—Cl，其中"—"象征共享的电子对，代表一个共价键。原子或原子团的化合价或其"钩子"数量简化了描述它需要捐赠、接受或共享多少电子以达到稀有气体电子结构的稳定性。以碳为例，其电子结构为 2.4，被认为具有四价，意味着它需要共享 4 个电子以达到氖的稳定构型 2.8。在某些情况下，原子会共享不止一对电子。例如，每个氧原子共享两对电子以达到氖的电子构型，用结构式 O=O 表示，表示一个双键。类似地，当三对电子被共享时，形成一个三键。

共价化合物的特点包括：

（1）挥发性及低熔点和沸点：这些化合物通常在室温下以液体或气体形态出现（例如，O_2、H_2O、N_2、CO_2），这是由于分子间缺乏强烈的静电吸引力，使它们能在较低温度下分离。将这些分子绑定在一起的弱引力称为范德瓦尔斯力。

（2）有限的水溶性：由于共价化合物结构中没有离子，它们与水分子的混合不易。然而，它们倾向于在非极性有机溶剂（如苯）中很好地溶解，这些溶剂也由共价分子组成。

（3）不导电：共价化合物中缺乏离子意味着它们不能导电，这与可以促进电流流动的离子化合物形成对比。

2.2.3 Metallic bonding

Metals are renowned for their unique properties, which any comprehensive bonding model must explain. One fundamental aspect of metals is their crystalline structure, which can be visualized as a rigid framework where particles align in a consistent, repeating arrangement. Within this framework, we can conceptualize a metallic crystal as composed of a lattice filled with positive ions. These ions are enveloped in an electron cloud, with electrons that freely traverse the structure. This setup facilitates an electrostatic attraction between the positive ions and the negatively charged electron cloud, effectively cementing the structure's integrity. In this process, metal atoms forfeit their outer (valence) electrons, transforming into positive ions. These relinquished electrons join the electron cloud, and due to their mobility, they cease to be associated with any specific atom from which they were liberated.

This model sheds light on the distinctive properties of metals:

(1) Reflectivity and Color: Metals are inherently shiny, a characteristic stemming from the free electrons' ability to absorb and subsequently re-emit incoming light. The vast majority of incident light on the metal surface is re-emitted, lending metals their lustrous appearance. This reflective quality is precisely why metals find widespread use in mirror production. While most metals exhibit a silvery hue, certain metals like copper and gold display red/gold colors. This divergence in coloration is due to these metals' selective absorption of specific visible light wavelengths over others.

(2) Electrical and Thermal Conductivity: The presence of free electrons enables metals to conduct electricity when in solid form, setting them apart from ionic compounds, which do not exhibit the same property. When electricity courses through a metal, there's no resultant chemical alteration of the metal itself. Applying a potential difference across a metal piece causes the electron cloud to migrate towards the positive potential. These electrons are also capable of rapidly transferring kinetic energy throughout the solid structure, elucidating why metals are excellent heat conductors.

(3) Malleability and Ductility: The overall structure remains intact, even when distorted, thanks to the cohesive force exerted by the electron cloud. This cohesive force

underpins metals' malleability and ductility—properties that allow metals to be shaped or drawn into thin wires without fracturing. The metallic bond's strength is directly proportional to the ratio of free electrons within the electron cloud to the number of ions. Consequently, the more valence electrons a metal atom contributes to the electron cloud, the more robust the resultant metallic bond.

New words and expressions

bonding model　结合模型
crystalline structure　结晶结构
positive ions　正离子
electron cloud　电子云
electrostatic attraction　静电吸引
reflectivity　反射性
thermal conductivity　热导率
potential difference　电势差
kinetic energy　动能
malleability　延展性
ductility　延性
metallic bond　金属键

译文

  金属因其独特的性质而闻名，任何全面的结合模型都必须解释这些性质。金属的一个基本特点是其结晶结构，可以视为一个刚性框架，其中粒子以一致、重复的排列对齐。在这个框架中，可以将金属晶体概念化为由填满正离子的晶格组成。这些离子被电子云包围，电子在结构中自由穿行。这种设置促进了正离子和带负电的电子云之间的静电吸引力，有效地巩固了结构的完整性。在这个过程中，金属原子放弃了它们的外层（价）电子，转化为正离子。这些放弃的电子加入了电子云，并由于它们的流动性，它们不再与任何特定的释放它们的原子相关联。

  这个模型揭示了金属的独特性质：

  （1）反射性和颜色：金属天生具有光泽，这是自由电子吸收并随后重新发射入射光所致。金属表面的绝大多数入射光被重新发射，使金属具有光亮的外观。这种反射品质正是金属在镜子生产中广泛使用的原因。虽然大多数金属表现为银白色，但某些金属如铜和金则显示红色或金色。这种颜色差异是这些金属选择性吸收特定的可见光而非其他波长的光所致。

  （2）电导率和热导率：自由电子的存在使金属在固态时能够导电，这使它们区别于不表现出同样属性的离子化合物。当电流通过金属时，金属本身不会发生化学变化。在金属片上施加电势差会导致电子云向正电势迁移。这些电子还能够在整个固体结构中迅速传递动能，这解释了为什么金属是优秀的热导体。

（3）延展性和延性：即使在变形时，由于电子云施加的凝聚力，整体结构仍然保持完整。这种凝聚力是金属延展性和延性的基础——这些属性使金属可以被塑造或拉成细丝而不会断裂。金属键的强度与电子云中自由电子与离子数量的比例直接成正比。因此，金属原子为电子云贡献的价电子越多，形成的金属键就越坚固。

2.3　Polyatomic ions

Both ionic and covalent bonding exist within some compounds. Consider the deadly poisonous compound potassium cyanide (KCN). This is an ionic compound; it contains the ions K^+ and CN^-. The atoms C and N in the cyanide ion achieve the stable electronic arrangement of inert gases. They do this by covalent bonding within the ion. The structural formula of the ion is written as $[C\equiv N]^-$. Ions such as CN^- that contain more than one atom are called polyatomic ions. Other examples of polyatomic anions that form ionic compounds, but have covalent bonding within the anion are hydroxide (OH^-), nitrate (NO_3^-), sulfate (SO_4^{2-}), carbonate (CO_3^{2-}), hydrogen carbonate (HCO_3^-) and phosphate (PO_4^{3-}).

Common ion roots are shown in Table 2.1.

表 2.1　常见离子根

Ions	Academic name	中文术语
NO_2^-	nitrite	亚硝酸根
NO_3^-	nitrate	硝酸根
SO_3^{2-}	sulfite	亚硫酸根
SO_4^{2-}	sulfate	硫酸根
HSO_3^-	bisulfite	亚硫酸氢根
HSO_4^-	bisulfate	硫酸氢根
ClO_2^-	chlorite	亚氯酸根
ClO_3^-	chlorate	氯酸根

2.3 Polyatomic ions

续表2.1

Ions	Academic name	中文术语
ClO^-	hypochlorite	次氯酸根
ClO_4^-	perchlorate	高氯酸根
NH_4^+	ammonium	铵离子
NH_3^+	ammonia	氨离子
OH^-	hydroxide	氢氧根
CN^-	cyanide	氰根
CO_3^{2-}	carbonate	碳酸根
$C_2O_4^{2-}$	oxalate	草酸根
O_2^{2-}	peroxide	过氧化物
PO_4^{3-}	phosphate	磷酸根
HCO_3^-	hydrogen carbonate	碳酸氢根
$H_2PO_4^-$	dihydrogen phosphate	磷酸二氢根
$C_2H_3O_2^-$	acetate	醋酸根
MnO_4^-	permanganate	高锰酸根
CrO_4^{2-}	chromate	铬酸根
$Cr_2O_7^{2-}$	dichromate	重铬酸根
H_3O^+	hydronium	水合氢离子
SCN^-	thiocyanate	硫氰酸根
$C_6H_5O_7^{3-}$	citrate	柠檬酸根

New words and expressions

compound   化合物
potassium cyanide   氰化钾
structural formula   结构式
polyatomic ions   多原子离子
hydroxide   氢氧根

nitrate   硝酸根
sulfate   硫酸根
carbonate   碳酸根
hydrogen carbonate   碳酸氢根
phosphate   磷酸根

译文

某些化合物中既存在离子键也存在共价键。考虑一下致命的毒性化合物氰化钾（KCN）。这是一种离子化合物，它包含离子 K^+ 和 CN^-。氰根离子中的 C 原子和 N 原子通过离子内的共价键实现了类似惰性气体的稳定电子排列。该离子的结构式写为 $[C \equiv N]^-$。像 CN^- 这样包含多于一个原子的离子称为多原子离子。其他形成离子化合物但阴离子内有共价键的多原子阴离子的例子包括氢氧根（OH^-）、硝酸根（NO_3^-）、硫酸根（SO_4^{2-}）、碳酸根（CO_3^{2-}）、碳酸氢根（HCO_3^-）和磷酸根（PO_4^{3-}）。

2.4 Allotropes

Carbon, a versatile element, manifests itself in three distinct solid forms known as allotropes. These allotropes showcase carbon atoms bonded in uniquely different configurations, demonstrating the element's capability to adopt various structural forms. Among these allotropes are diamond, graphite, and buckminsterfullerene, each with its own set of properties and uses derived from the particular bonding arrangements of the carbon atoms.

Diamond, the first of these allotropes, is celebrated for its brilliance and extraordinary hardness. It stands out as a transparent solid, notable for its non-conductive nature regarding electricity. The structural integrity of diamond comes from each carbon atom forging single covalent bonds with four others, positioned at the tetrahedron's corners. This extensive network of strong carbon-carbon single bonds pervades the entire crystal, forming a network solid where the atoms are cohesively bonded throughout the entire structure. Such a robust interconnectedness renders diamond incredibly hard, necessitating extremely high temperatures to disrupt its lattice for melting. Diamond's remarkable melting point, soaring to 3823 K, is a testament to the strength and resilience of its bonding.

Graphite, on the other hand, offers a stark contrast to diamond's characteristics. It is a soft, black solid that possesses the rare ability to conduct electricity, a trait not typically associated with covalently bonded materials. Graphite's structure comprises layers of carbon atoms where each atom is covalently bonded to three others in a trigonal planar fashion. Although the bonds within each layer are strong, the layers themselves are interconnected by relatively weaker London dispersion forces, allowing them to slide over one another with ease. This unique structural feature gives graphite its slippery texture, making it an ideal material for lubricants and the familiar "lead" in pencils. The ease with which graphite marks paper is due to the facile shearing of these layers. Moreover, the presence of a cloud of free electrons, reminiscent of those in metallic bonds, enables graphite's electrical conductivity and contributes to its lustrous appearance. Graphite's microscopic form, soot, is composed of small graphite crystals.

Buckminsterfullerene, unlike its ancient counterparts diamond and graphite, is a relatively recent discovery made in 1985. It does not fit the traditional mold of a giant molecule as diamond and graphite do, yet its molecular structure is significantly larger than typical molecular solids such as S_8 or P_4. Named after American architect Buckminster Fuller, whose geodesic domes the molecule's structure mimics, C_{60} also bears resemblance to a soccer ball with its interlocking hexagons and pentagons. This intriguing structure has led to the informal moniker "buckyball" for this allotrope. At room temperature, buckminsterfullerene takes the form of a dark solid and is soluble in covalent solvents like benzene. The potential applications for buckminsterfullerene and its chemical derivatives are vast, spanning from lubricants to batteries and semiconductors. The discovery of C_{60} was so groundbreaking that it earned Harry Kroto, Richard Smalley, and Robert Curl the 1996 Nobel Prize in Chemistry, highlighting the significance of this carbon allotrope in the scientific community.

New words and expressions

tetrahedron 四面体	London dispersion forces 伦敦色散力
network solid 网络固体	lubricants 润滑剂
crystal 晶体	conduct electricity 导电
layers 层	graphite crystals 石墨晶体
allotropes 同素异形体	Buckminsterfullerene 巴克敏斯特富勒烯
trigonal planar arrangement 三角平面排列	molecular solids 分子固体

soluble	可溶的	Nobel Prize in Chemistry	诺贝尔化学奖
chemical derivatives	化学衍生物		
semiconductors	半导体		

译文

碳这一多才多艺的元素以 3 种不同的固态形式表现出来，这些形式被称为同素异形体。这些同素异形体展示了以独特的配置相连的碳原子，展现了元素采用各种结构形态的能力。这些同素异形体包括钻石、石墨和巴克敏斯特富勒烯，每种都因其碳原子的特定键合排列而拥有一套独特的属性和用途。

钻石，这些同素异形体中的第一个，因其光辉和非凡的硬度而受到赞誉。它是一个透明固体，以其不导电的特性而著名。钻石的结构完整性来源于每个碳原子与其他 4 个碳原子形成单个共价键，这些碳原子位于四面体的角落。这个广泛的强碳—碳单键网络贯穿整个晶体，形成了一个网络固体，其中的原子在整个结构中紧密相连。这种强大的互连使钻石异常硬，需要极高的温度才能破坏其晶格结构使其熔化。钻石的熔点令人难以置信，高达 3823 K，证明了其键合的强度和韧性。

另外，石墨与钻石的特性形成鲜明对比。它是一种软的黑色固体，拥有罕见的导电能力，这是一个不典型的与共价键合物质相关的特性。石墨的结构由碳原子层组成，每个碳原子在三角平面排列中与其他 3 个碳原子共价键合。尽管每层内的键很强，但层与层之间通过相对较弱的伦敦色散力相互连接，使得它们能够轻易地滑动。这一独特的结构特性赋予了石墨滑腻的质感，使其成为理想的润滑剂和铅笔中熟悉的"铅"材料。石墨之所以能在纸上留下痕迹，是因为这些层很容易被擦掉。此外，类似于金属键中的自由电子云使得石墨能够导电并赋予其光泽外观。石墨的微观形态，烟灰，由小的石墨晶体组成。

与其古老的对应物钻石和石墨不同，巴克敏斯特富勒烯是在相对较近期的 1985 年发现的。它不像钻石和石墨那样符合巨分子的传统模式，然而其分子结构明显大于典型的分子固体如 S_8 或 P_4。以美国建筑师巴克敏斯特·富勒命名，其分子结构模仿了富勒的地质穹顶，C_{60} 也类似于一个足球，由相互连接的六边形和五边形组成。这种有趣的结构使得这种同素异形体被非正式地称为"巴基球"。在室温下，巴克敏斯特富勒烯呈现为一种深色固体，并且可以溶解于共价溶剂如苯中。巴克敏斯特富勒烯及其化学衍生物的潜在应用非常广泛，从润滑剂到电池和半导体等。C_{60}的发现如此具有开创性，以至于哈里·克罗托、理查德·斯莫利和罗伯特·卡尔1996年被授予诺贝尔化学奖，凸显了这种碳同素异形体在科学界的重要性。

2.5 Soluble and insoluble

2.5.1 Solution

In Chemistry, a solution is a homogeneous mixture made up of two components: the solvent, which is the medium that dissolves the other component, and the solute, which is the substance being dissolved. Taking the dissolution of sugar in water as an illustration, water serves as the solvent, sugar acts as the solute, and together they form a sugary solution. Continually incorporating sugar into the water will eventually lead to a saturation point where the water can no longer dissolve additional sugar, rendering the solution saturated. At this juncture, any excess sugar added merely accumulates at the container's base. Interestingly, elevating the solution's temperature increases the water's capacity to dissolve more sugar before reaching saturation. This phenomenon, where many solids including sugar dissolve more readily at elevated temperatures, contrasts with gases, which exhibit reduced solubility in warm water compared to cold.

The principles of solubility hinge on the nature of the substances involved. Polar substances, characterized by the presence of ions or polar molecules, dissolve exclusively in polar solvents, which themselves comprise polar molecules. Conversely, non-polar substances dissolve solely in non-polar solvents. This concept is encapsulated in the adage "like dissolves like", indicating that substances dissolve in solvents of a similar polarity. By assessing their mutual solubility, solvents can be arranged on a polarity scale. Among commonly used solvents, water ranks as the most polar, whereas hydrocarbons such as heptane and hexane are on the lower end of the polarity spectrum.

Regarding miscibility, when two solvents mix to form a single homogeneous layer, they are considered miscible. If the mixture results in two distinct layers, each composed of the pure solvents without mixing, the solvents are deemed immiscible. Should two layers form, the less dense solvent will float atop the denser one. In this context, "layer" can be interchangeably used with "phase", exemplified by a hexane and water mixture, which separates into two phases due to their immiscibility.

The concept of partial miscibility reveals that very few solvents are completely immiscible. Even when two liquids seem to separate distinctly, a minuscule quantity of each solvent will still be present in the opposing layer. For instance, mixing 100 grams of hexane with 100 grams of water at room temperature results in nearly immiscible layers. However, the hexane layer will contain a small amount of water (approximately 0.01 grams), and similarly, the water layer will have a trace amount of hexane (about 0.001 grams), illustrating the nuanced interplay of solubility and miscibility in chemical solutions.

New words and expressions

solution　溶液
solvent　溶剂
solute　溶质
dissolve　溶解
saturation point　饱和点
saturated　饱和的
temperature　温度
polar substance　极性物质
polar molecules　极性分子

polar solvents　极性溶剂
non-polar substance　非极性物质
non-polar solvents　非极性溶剂
miscibility　互溶性
immiscible　不互溶
density　密度
phase　相
partially miscible solvents
　　部分互溶溶剂

译文

在化学中，溶液是由两个组分组成的均匀混合物：溶剂，作为溶解其他组分的介质；以及溶质，即被溶解的物质。以糖在水中溶解为例，水充当溶剂，糖则是溶质，二者共同形成糖溶液。不断向水中加入糖，最终会达到一个饱和点，此时水无法再溶解更多的糖，使溶液达到饱和状态。若再添加糖只会积累在容器的底部。有趣的是，提高溶液的温度可以增加水在达到饱和之前能溶解的糖的量。这种现象表明，许多固体包括糖在高温下的溶解度更高，而与之相反，气体在热水中的溶解度比在冷水中低。

溶解度的原理依赖于涉及物质的性质。含有离子或极性分子的极性物质只能在极性溶剂中溶解，而这些极性溶剂本身由极性分子组成。相反，非极性物质只能在非极性溶剂中溶解。这一概念被概括为"相似溶于相似"的规则，表明物质可在与其极性相似的溶剂中溶解。通过评估它们之间的相互溶解性，可以按极性排序溶剂。在常用溶剂中，水是最极性的，而如庚烷和己烷这样的烃类是最不极性的。

关于互溶性，当两种溶剂混合形成单一均匀层时，称这些溶剂为互溶的。如果混合结果形成两个独立层，且每层都由纯溶剂组成，则这些液体被认为是不互溶的。如果形成两层，密度较低的溶剂会漂浮在上层。此时，"层"这个词经常被"相"这个词替换。因此，己烷和水的混合物会形成两个相，因为它们不互溶。

对于部分互溶溶剂，很少有溶剂是完全不互溶的，即使两种液体看起来不混合，每种溶剂在另一层中仍会有少量存在。例如，假设在室温下将 100 g 己烷与 100 g 水混合。这些液体几乎不互溶，但己烷层实际上会含有约 0.01 g 的水，而水层会含有约 0.001 g 的己烷，这展示了溶解度和互溶性在化学溶液中的微妙相互作用。

2.5.2　Solubility

Defining the solubility of ionic substances can be approached in two convenient ways: either by specifying the amount of substance in grams per 1000 grams of a saturated solution or by indicating the moles of substance required to create 1 dm^3 (liter) of a saturated solution. Molar solubility, which is denoted by the symbol (S), describes the concentration of a compound (in mol/dm^3) within a saturated solution at a specific temperature.

Ionic compounds that exhibit limited solubility in water are categorized as sparingly soluble. Notable examples of such compounds include silver chloride (AgCl) and barium sulfate ($BaSO_4$), both of which dissociate into their respective ions (Ag^+ and Cl^- for silver chloride, Ba^{2+} and SO_4^{2-} for barium sulfate) in solution. Given the minimal solubility of these compounds, the volume of water utilized to dissolve a given amount of the salt can generally be assumed to be equivalent to the total volume of the resulting solution. For instance, dissolving 0.0010 grams of silver chloride in 1000 cm^3 of water permits the assumption that the solution's volume remains at 1000 cm^3. However, this approximation is not valid for highly soluble salts, as significant concentrations of the salt can slightly alter the volume of the liquid.

It's crucial to remember that while many ionic substances have limited solubility in water, those that do dissolve undergo complete dissociation into their constituent ions, a process fundamental to understanding the behavior of ionic compounds in solution.

New words and expressions

ionic substances　离子物质
saturated solution　饱和溶液
moles　摩尔
molar solubility　摩尔溶解度
concentration　浓度
sparingly soluble　微溶的
silver chloride　氯化银
barium sulfate　硫酸钡
dissociates　离解
separate ions　分离的离子

译文

定义离子物质的溶解度时，有两种便捷的方法：一种是指定每 1000 g 饱和溶液中的物质的质量（以克为单位），另一种是指出生成 1 dm³（升）饱和溶液所需的物质的摩尔数。摩尔溶解度，用符号 S 表示，描述了在特定温度下饱和溶液中化合物的浓度（mol/L）。

在水中溶解度有限的离子化合物被称为微溶物。这类化合物的典型例子包括氯化银（AgCl）和硫酸钡（$BaSO_4$），它们在溶液中分离成相应的离子（氯化银分离成 Ag^+ 和 Cl^-，硫酸钡分离成 Ba^{2+} 和 SO_4^{2-}）。鉴于这些化合物的溶解度很低，用于溶解一定量盐的水体积通常可以假定等于最终溶液的总体积。例如，将 0.0010 g 氯化银溶解在 1000 cm³ 的水中时，可以安全地假定溶液的体积也为 1000 cm³。然而，这种近似对于高溶解度的盐不适用，因为盐的显著浓度会轻微改变液体的体积。

重要的一点是，虽然许多离子物质在水中的溶解度有限，但那些确实溶解的物质会完全离解成它们的组成离子，这一过程是理解离子化合物在溶液中行为的基础。

2.5.3　Insoluble

Colloids

Colloids, an intriguing category of mixtures, occupy the middle ground between true solutions, like sugar dissolved in water, and suspensions, such as chalk or sand mixed with water. This group encompasses a wide array of substances, including tea, starch, and milk when mixed with water, as well as gelatine (a type of protein) in water, the semi-solid aluminium hydroxide gel, and soapy water. These mixtures are often referred to by several terms, including colloidal solutions, colloidal suspensions, or colloidal dispersions, highlighting their diverse nature and applications.

Distinguishing features of colloids

(1) Particle size: The defining characteristic of colloids is the size of their particles, which typically range from 1 to 1000 nanometers in diameter. These dimensions render the particles invisible under traditional optical microscopes. Colloidal particles might be either large molecules, such as proteins, or aggregates comprising multiple molecules or ions.

(2) Behavior: Unlike the particles in suspensions, colloidal particles don't settle at the container's bottom, nor can they be filtered out using standard filter paper. Instead, a process known as dialysis, which involves the use of a special membrane, is required to separate colloids from true solutions.

(3) Optical properties: A distinctive way to identify colloids from true solutions involves the Tyndall effect, observed when light is passed through the colloid. The relatively large size of the colloidal particles scatters light, a phenomenon that can be likened to the scattering of car headlights in fog.

Emulsions: a special type of colloid

Emulsions are specific colloidal dispersions where tiny droplets of one liquid are distributed throughout another. Common in various commercial products, foods, and biological systems, emulsions require vigorous mixing to achieve temporary dispersion, such as shaking oil and water together. However, for permanent dispersion, the addition of an emulsifying agent is necessary.

Milk serves as a natural emulsion of butterfat droplets in a water-based solution containing sugars, mineral salts, and proteins. Similarly, ice cream is another emulsion, with proteins and smaller amounts of phospholipids like lecithin acting as the emulsifying agents. In egg yolk, phospholipids and cholesterol fulfill this role.

Emulsions significantly aid the digestive process, especially the breakdown of fats in the intestines. By emulsifying fats into smaller droplets, the surface area for enzymatic action increases, thereby accelerating fat digestion.

Soap: a colloidal emulsifier

Soap, along with other detergents, forms colloidal particles known as ionic micelles. These micelles consist of 50 to 100 ions bonded together and play a crucial role in the emulsification of fats and oils, thus facilitating their removal. Beyond their cleaning capabilities, soaps also find applications as emulsifying agents in numerous cosmetic products, underscoring their versatility in both household and industrial settings.

New words and expressions

colloids	胶体	optical microscopes	光学显微镜
true solutions	真溶液	dialysis	透析
suspensions	悬浮液	tyndall effect	廷德尔效应
gelatine	明胶	emulsions	乳状液
aluminium hydroxide gel	氢氧化铝凝胶	emulsifying agent	乳化剂
		butterfat droplets	黄油脂滴
soapy water	肥皂水	phospholipids	磷脂
colloidal solutions	胶体溶液	lecithin	卵磷脂
colloidal suspensions	胶体悬浮液	ionic micelles	离子胶束
colloidal dispersions	胶体分散体	detergents	洗涤剂
particle size	粒子大小	cosmetics	化妆品

译文

胶体及其特性

胶体是一种独特的混合物类型，介于真溶液（如水中的糖）和悬浮液（如水中的粉笔或沙子）之间。这一类包括的物质范围广，例如水中的茶、淀粉和牛奶，以及水中的明胶（一种蛋白质）、氢氧化铝凝胶和肥皂水。这些混合物通常被称为胶体溶液、胶体悬浮液或胶体分散体，凸显了它们的多样性和应用。

胶体的区别特征

（1）粒子大小：胶体的定义特征是其粒子的大小，通常在 1~1000 nm。这些尺寸使得粒子在传统的光学显微镜下不可见。胶体粒子可能是大分子（如蛋白质）或分子或离子的聚集体。

（2）行为：与悬浮液中的粒子不同，胶体粒子不会以同样的方式沉积在容器底部，也不能使用普通滤纸过滤掉。然而，胶体可以使用特殊的膜通过透析过程从真溶液中分离出来。

（3）光学性质：通过照射光线穿过溶液来区分胶体和真溶液是一种独特的方法。相对较大的胶体粒子会散射光线，这被称为廷德尔效应。汽车头灯在雾中的散射是这种效应的一个熟悉例子。

乳状液：一种特殊类型的胶体

乳状液是一种特定的胶体分散体，其中一种液体以微小滴的形式分布在另一种液体中。乳状液在各种商业产品、食品和生物系统中很常见，需要通过剧烈混合来实现临时分散，例如将油和水一起摇晃。然而，为了实现永久分散，需要添加乳化剂。

牛奶是黄油脂滴在含有糖、矿物盐和蛋白质的水溶液中的自然乳状液。类似地，冰激凌也是一种乳状液，其乳化剂是蛋白质和较少量的磷脂，主要是卵磷脂。在蛋黄中，乳化剂是磷脂和胆固醇。

乳状液在肠道中脂肪的消化过程中扮演着重要角色。在小肠中，脂肪被胰腺和肝脏的碱性分泌物乳化。为什么在脂肪能被消化之前需要乳化它们？答案是脂肪被消化的速率取决于暴露于肠道的脂肪"溶液"的表面积。例如，如果将 1 cm^3 的油分散成直径为 5 nm 的油滴，所有滴的总面积约为 1200 m^2。乳化能如此有效地展开油滴，以至于酶可以更快地分解脂肪。

肥皂：一种胶体乳化剂肥

皂和其他洗涤剂形成的胶体粒子称为离子胶束。这些胶束通常包含 50~100 个黏合在一起的离子，对于解释肥皂和洗涤剂如何乳化脂肪和油脂至关重要。除了用于洗涤衣物外，肥皂还在许多化妆品中用作乳化剂，凸显了它们在家庭和工业设置中的多功能性。

2.6　Common organic compounds

A plethora of organic chemicals originate from hydrocarbons, which can be transformed by substituting one or several hydrogen atoms with different atoms or groups, known as functional groups. This method introduces a way to categorize the immense array of organic compounds into distinct "families". Each family is unified by common functional groups, and the characteristic reactions of these groups define the family's properties. Familiarizing oneself with these functional groups is a crucial step before delving deeper into the subject.

One notable family is the Halogenoalkanes, also known as alkyl halides. These compounds are synthesized by replacing a hydrogen atom in a hydrocarbon with a halogen element, such as Fluorine (F), Chlorine (Cl), Bromine (Br), or Iodine (I). This substitution results in compounds with a general formula of R—X, where "R" represents the hydrocarbon portion and "X" stands for the halogen. Examples include CH_3F (fluoromethane), CH_3Cl (chloromethane), CH_3Br (bromomethane), and CH_3I (iodomethane). Notably, Halogenoalkanes do not mix well with water, leading to the formation of two distinct layers when combined; the denser halogenoalkane layer often settles below the water layer.

Among the halogenoalkanes, certain compounds have gained prominence for their uses and hazards. Trichloromethane, also known as chloroform ($CHCl_3$), was once a popular anaesthetic due to its non-flammable nature, until its potential to cause liver damage was discovered. Tetrachloromethane, or carbon tetrachloride (CCl_4), shares similar anaesthetic properties but is recognized for its greater toxicity. Bromochlorotrifluoroethane, commonly referred to as halothane, has emerged as a widely accepted alternative. It retains the anaesthetic benefits while presenting a safer option for use.

Alcohol Compounds: These substances are characterized by the formula R—OH, where the "R" represents an alkyl group attached to a hydroxyl (—OH) group. The nomenclature of alcohols concludes with the suffix "ol", and in cases where the position of the hydroxyl group on the carbon chain is significant, a numerical prefix directly before "ol" identifies the specific carbon bonded to the —OH group. For alcohols featuring multiple —OH groups, terms such as diol, triol, etc., are used to denote the count of such hydroxyl groups, with compounds like 1,2-diol also known as glycols due to their structure.

The boiling points of alcohols exceed those of other organic molecules with comparable molecular weights due to their capacity for forming hydrogen bonds among themselves. This hydrogen bonding capability also endows lower molecular weight alcohols with solubility in water, making them miscible. Ethane-1,2-diol, commonly known as ethylene glycol, exemplifies this as it readily dissolves in water. Its utility in antifreeze stems from its dual hydroxyl groups that enable extensive hydrogen bonding with water, alongside its notable boiling point and minimal freezing point.

Solvent properties of alcohols: The hydroxyl group in alcohols lends them partial water-like solvent characteristics, facilitating the dissolution of ionic compounds. Moreover, the alkyl portion (R-group) allows them to dissolve organic, covalent materials, making alcohols versatile solvents.

Ethanol specifics: Among the myriad of alcohols, ethanol stands out as the one most commonly associated with the term "alcohol" in everyday language. The creation of ethanol is primarily through the fermentation of carbohydrates, a natural process involving the breakdown of sugars. For instance, grape juice, rich in the sugar glucose ($C_6H_{12}O_6$), can transform into ethanol with the introduction of yeast. Yeast, a fungal organism, produces the enzyme zymase, catalyzing the conversion of glucose to ethanol. Enzymes, as biological catalysts, facilitate this transformation, optimally conducted around 20 ℃ to maintain yeast viability and ensure efficient fermentation.

Fermentation process and ethanol production: The fermentation process halts when the ethanol concentration in the mixture reaches approximately 14% by volume. At this concentration, ethanol is lethal to the yeast cells responsible for fermentation. To obtain alcohol with a higher ethanol content, the solution undergoes fractional distillation. Alcoholic beverages like whisky or brandy typically have an ethanol content around 40%. Although fermentation is a traditional method for producing alcoholic drinks, it's not the preferred method for manufacturing ethanol in bulk for chemical purposes. Ethanol can also be synthesized on an industrial scale through the hydration of alkenes, where water is added across the double bond of an alkene, resulting in the formation of an alcohol.

Effects of ethanol consumption: Consuming diluted ethanol solutions, as found in alcoholic beverages, can induce feelings of relaxation and pleasure. However, excessive consumption leads to adverse effects such as headaches and vomiting, and chronic misuse can cause severe liver damage.

Methanol and its uses: Historically, methanol, or wood alcohol, was obtained by distilling wood. Modern methods produce methanol through the catalytic hydrogenation of carbon monoxide. Methanol serves as a solvent and an antifreeze in car radiators but is significantly more toxic than ethanol; ingestion can cause blindness or even death. Methanol also finds use as a fuel for vehicles.

Methylated spirits: Due to the high taxation on alcoholic beverages in many regions, ethanol's extensive industrial applications—ranging from the production of organic chemicals to serving as a vital industrial solvent—necessitate a form of ethanol that is exempt from such taxes. To this end, ethanol is denatured, typically by adding methanol, creating "industrial methylated spirit". For household use, this product is often dyed purple to signal its toxicity and prevent accidental ingestion.

Optical isomerism in organic molecules: Certain organic molecules can exist as two forms that are mirror images of each other, akin to a pair of gloves. This phenomenon occurs when a carbon atom is bonded to four distinct atoms or groups, resulting in two non-superimposable isomers, with the carbon atom referred to as a chiral centre. These mirror image isomers, known as enantiomers, share similar chemical and physical properties but differ in their interaction with plane-polarized light, being optically active and rotating light in opposite directions.

Amino acids: Amino acids, the building blocks of proteins, contain both an amino group ($-NH_2$) and a carboxylic acid group ($-COOH$). These compounds are non-volatile, crystalline solids that are highly soluble in water and often optically active. The presence of both acidic and basic groups in amino acids allows for the formation of a dipolar ion or zwitterion in solution, imparting salt-like properties to the molecule.

Peptides and proteins: Peptides form through the reaction between the amino group of one amino acid and the carboxyl group of another, creating a peptide bond (—CONH—). The terms dipeptides and tripeptides refer to molecules derived from two or three amino acids, respectively, while polypeptides describe chains with a molecular mass up to 10,000 u. Molecules exceeding this size are classified as proteins, which are essential components of all living cells and play various roles, such as in enzymes, hormones, and structural elements like keratin in nails and hair. Proteins are categorized into fibrous proteins, which are long, tough, and water-insoluble, and globular proteins, which are compactly folded molecules found in enzymes, hormones, and hemoglobin.

New words and expressions

organic chemicals　有机化学品
hydrocarbons　烃
hydrogen atoms　氢原子
functional groups　功能团
halogenoalkanes (alkyl halides)　卤代烷 (烷基卤化物)
chlorine (Cl)　氯
trichloromethane (chloroform)　三氯甲烷 (氯仿)
tetrachloromethane (carbon tetrachloride)　四氯甲烷 (四氯化碳)
bromochlorotrifluoroethane (halothane)　溴氯三氟乙烷 (哈罗烷)
alcohol compounds　醇类化合物
nomenclature　命名
numerical　数字的
prefix　前缀
diol　二醇
triol　三醇
glycol　乙二醇
hydroxyl group　羟基
ethylene glycol　乙二醇
enzymes　酶
fermentation　发酵
fractional distillation　分馏
hydration of alkenes　烯烃的水合
methanol　甲醇
methylated spirits　甲醇精
alkyl group　烷基
carboxyl group　羧基
ethanol　乙醇
antifreeze　防冻剂
industrial solvent　工业溶剂
optical isomerism　光学异构
organic molecules　有机分子
carbon atom　碳原子
chiral centre　手性中心
enantiomers　对映体

plane-polarized light　平面偏振光
amino acids　氨基酸
amino group　氨基
carboxylic acid group　羧酸基
crystalline solids　结晶性固体
dipolar ion　偶极离子
zwitterion　孑体离子
peptides　多肽
peptide bond　肽键

dipeptides　二肽
tripeptides　三肽
polypeptides　多肽链
proteins　蛋白质
fibrous proteins　纤维蛋白
globular proteins　球蛋白
hormones　激素
keratin　角蛋白
hemoglobin　血红蛋白

译文

大量的有机化学品来源于烃，这些烃可以通过用不同的原子或基团（称为功能团）替换一个或多个氢原子来转化。这种方法引入了一种将庞大的有机化合物数组分类为不同"家族"的方式。每个家族通过共有的功能团统一，并且这些功能团的特征反应定义了家族的属性。在深入探讨这一主题之前，熟悉这些功能团是至关重要的一步。

一个值得注意的家族是卤代烷，也称为烷基卤化物。这些化合物通过用卤素元素（如氟、氯、溴或碘）替换烃中的一个氢原子来合成。这种替换产生了具有通用公式 R—X 的化合物，其中 R 代表烃部分，X 代表卤素。例子包括 CH_3F（氟甲烷）、CH_3Cl（氯甲烷）、CH_3Br（溴甲烷）和 CH_3I（碘甲烷）。值得注意的是，卤代烷与水混合不良，混合时会形成两个明显的层次，密度较大的卤代烷层常常沉于水层之下。

在卤代烷中，某些化合物因其用途和危害而脱颖而出。三氯甲烷，也称为氯仿（$CHCl_3$），曾因其不易燃而作为一种流行的麻醉剂，直到人们发现其可能导致肝脏损伤。四氯甲烷或四氯化碳（CCl_4）具有类似的麻醉属性，但因其更大的毒性而被认识。溴氯三氟乙烷，通常称为哈罗烷，已成为被广泛接受的替代品。它保留了麻醉的好处，同时提供了一个更安全的使用选项。

醇类化合物：这些物质的特点是公式 R—OH，其中 R 代表附着一个羟基（—OH）的烷基。醇的命名以"ol"作为后缀结尾，如果羟基在碳链上的位置很重要，直接在"ol"之前的数字前缀标识与—OH 基团相连的特定碳。对于具有多个—OH 基团的醇，如二醇、三醇等术语用于表示这种羟基的数量，具有 1,2-

二醇结构的化合物也被称为乙二醇，因其结构而得名。

醇的沸点由于它们能够在自身间形成氢键而超过了具有可比分子量的其他有机分子。这种氢键能力也赋予了低分子量醇在水中的溶解性，使它们可混合。乙二醇，通常称为乙二醇，就是一个例子，它容易溶于水。它在防冻剂中的应用源于其双羟基使得与水的广泛氢键成为可能，以及其显著的沸点和极低的冰点。醇的溶剂属性：醇中的羟基使它们部分具有类似水的溶剂特性，促进了离子化合物的溶解。此外，烷基部分（R-组）允许它们溶解有机的共价材料，使醇成为多功能溶剂。

乙醇的特点：在众多醇类中，乙醇是最常与日常语言中的"醇"一词关联的。乙醇主要通过碳水化合物的发酵产生，这是一个自然过程，涉及糖的分解。例如，富含葡萄糖（$C_6H_{12}O_6$）的葡萄汁在加入酵母后可以转化为乙醇。酵母，一种真菌生物，产生酶，催化葡萄糖转化为乙醇的过程。作为生物催化剂的酶促进了这种转化，最佳进行温度约为20℃，以保持酵母的活力并确保发酵效率。

发酵过程和乙醇生产：当混合物中乙醇浓度达到约14%体积时，发酵过程停止。在这个浓度下，乙醇对负责发酵的酵母细胞是致命的。要获得更高乙醇含量的酒精，溶液需进行分馏。如威士忌或白兰地等酒精饮料的乙醇含量通常约为40%。虽然发酵是生产酒精饮料的传统方法，但它不是化学目的大批量制造乙醇的首选方法。乙醇也可以通过烯烃的水合在工业规模上合成，其中水添加到烯烃的双键上，产生醇。

乙醇消费的影响：消费稀释的乙醇溶液，如酒精饮料中所含，可以引起放松和愉悦的感觉。然而，过量消费会导致不良影响，如头痛和呕吐，长期滥用可以造成严重的肝脏损害。

甲醇及其用途：历史上，甲醇或木醇是通过蒸馏木材获得的。现代方法通过碳一氧化物的催化加氢产生甲醇。甲醇作为溶剂和汽车散热器中的防冻剂，但其毒性明显高于乙醇；摄入可能导致失明甚至死亡。甲醇也用作车辆的燃料。

甲醇精：由于许多地区对酒精饮料征收高额税款，乙醇在从生产有机化学品到作为至关重要的工业溶剂等方面的广泛工业应用，需要一种免税的乙醇形式。为此，乙醇被变性，通常通过添加甲醇，创建了"工业甲醇精"。对于家庭用途，这种产品经常被染成紫色，以示其毒性并防止意外摄入。

有机分子中的光学异构现象：某些有机分子可以存在为两种彼此为镜像的形式，类似于一副手套。当一个碳原子与4个不同的原子或基团相连时，会产生这种现象，导致两个不可叠加的异构体，这个碳原子被称为手性中心。这些镜像异构体，称为对映体，共享类似的化学和物理属性，但在与平面偏振光的相互作用上有所不同，它们是光学活性的，并且以相反的方向旋转光线。

氨基酸：氨基酸是蛋白质的构建块，含有一个氨基（—NH_2）和一个羧酸基（—COOH）。这些化合物是非挥发性的、结晶性固体，它们在水中高度可溶，并且经常光学活性。氨基酸中酸性和碱性基团的存在允许在溶液中形成偶极离子或子体离子，赋予分子类盐的性质。

多肽和蛋白质：多肽通过一个氨基酸的氨基与另一个氨基酸的羧基反应形成，创建了一个肽键（—CONH—）。二肽和三肽这些术语分别指从两个或三个氨基酸派生的分子，而多肽链描述的是分子量高达 10,000 u 的链。超过这个大小的分子被归类为蛋白质，它们是所有活细胞的必要组成部分，并扮演各种角色，如作为酶、激素，以及如指甲和头发中的角蛋白等结构元素。蛋白质被分类为纤维蛋白，它们是长的、坚韧的、不溶于水的，以及球蛋白，它们是在酶、激素和血红蛋白中发现的紧凑折叠的分子。

Chapter 3   Experimental Methods

3.1   Glassware

In a laboratory setting, various glassware and instruments are essential for conducting scientific research. Typically, the glassware encountered falls into three primary categories: beakers, flasks, and cylinders, each with distinct characteristics and applications.

Beakers, characterized by their straight parallel sides, resemble drinking cups but are not used for consumption purposes. Despite featuring numerical graduations along their sides, these markings are not intended for precise measurements but rather provide a rough estimate of volume. Beakers, lacking accuracy for meticulous volume measurements, are primarily utilized for holding substances. They are available in a range of sizes, with 250 milliliters being a common capacity, though they can vary significantly in size from much larger to as small as 10 milliliters.

Flasks, particularly the Erlènmeyer flasks, diverge from beakers with their conical shape, having sides that taper inward towards the narrow opening at the top. This design minimizes the risk of splashing, making flasks suitable for chemical reactions. The Erlenmeyer flask, named after its inventor, is frequently used in laboratories, with sizes typically starting at 25 milliliters.

For accurate liquid volume measurements, the graduated cylinder is the instrument of choice. It features detailed graduations along its side for precise volume assessments. Graduated cylinders come in various sizes, with narrower designs facilitating enhanced accuracy in measurements.

Test tubes, another ubiquitous piece of glassware, are used for conducting small-scale reactions. They are stored vertically in racks due to their rounded bottoms, which prevent them from standing upright on flat surfaces. Specialized equipment, such as test tube holders, facilitates safe handling, especially when heating is involved.

Crucible tongs, designed with unique ends for gripping, are used to handle hot items directly, such as crucibles and hot glassware. Beaker tongs, equipped with non-slip rubber grips, are preferred for handling larger, hot beakers.

Heating in the laboratory is commonly achieved with a Bunsen burner, capable of producing a flame that can reach temperatures around 700 degrees Celsius. For supporting and heating glassware, a ring stand equipped with iron rings or wire gauze is used. A clay triangle may also be employed for securely holding crucibles over the Bunsen burner.

For transferring solid powders or crystals, a scoopula serves as a hybrid scoop and spatula. Observations of substances can be made on a watch glass, though it is not suitable for heating due to its fragile nature. For evaporating liquids, an evaporating dish, made of durable porcelain, is used.

Filtration processes utilize a filter funnel, where a mixture is poured through filter paper to separate solids from liquids. Stirring, particularly in acidic or metal-reactive solutions, is performed with a glass stirring rod to avoid contamination. For liquid transfers, pipettes function similarly to eyedroppers but are disposable due to cleaning difficulties.

Temperature measurements in the laboratory are conducted with thermometers, typically alcohol-based for safety and breakage prevention, as opposed to mercury thermometers.

Understanding and familiarity with laboratory equipment, including their names and functions, are crucial for efficient and safe research practices. This knowledge is foundational for anyone working within a laboratory environment.

New words and expressions

beakers	烧杯	numerical graduations	数字刻度
flasks	烧瓶	side	边
cylinders	量筒	milliliter	毫升
Erlenmeyer flasks	埃伦迈耶烧瓶	graduated cylinder	刻度量筒

test tubes 试管	porcelain 陶瓷
test tube holders 试管架	watch glass 表面玻璃
crucible tongs 坩埚钳	evaporating dish 蒸发皿
beaker tongs 烧杯钳	funnel 漏斗
bunsen burner 本生灯	filtration 过滤
ring stand 铁环架	filter 过滤器
wire gauze 铁丝网	glass stirring rod 玻璃搅拌棒
clay triangle 黏土三角	pipettes 移液管
scoopula 药匙	thermometers 温度计
spatula 铲子	

译文

在实验室环境中，各种玻璃器皿和仪器对于进行科学研究至关重要。通常遇到的玻璃器皿分为三个主要类别：烧杯、烧瓶和量筒，每种都有其独有的特征和应用。

烧杯以其直的平行边缘为特征，看起来像饮用杯，但不用于饮用目的。尽管其侧面有数值刻度，这些标记不是用于精确测量，而是提供体积的大致估计。由于缺乏精确体积测量的准确性，烧杯主要用于容纳物质。它们有多种尺寸，常见的容量为 250 mL，尽管大小可以从更大到小至 10 mL 不等。

烧瓶，特别是埃伦迈耶烧瓶，与烧杯不同，它们呈锥形，侧面向顶部的窄口逐渐收窄。这种设计最小化了溅出的风险，使烧瓶适合进行化学反应。以其发明者命名的埃伦迈耶烧瓶在实验室中经常使用，大小通常从 25 mL 开始。

对于精确的液体体积测量，刻度量筒是首选仪器。它的侧面有详细的刻度，用于精确的体积评估。刻度量筒有各种尺寸，较窄的设计有助于提高测量的准确性。

试管是另一种普遍使用的玻璃器皿，用于进行小规模反应。由于它们的底部是圆的，所以它们在架子上垂直存放，防止它们在平坦的表面上直立。专门的设备，如试管架，便于安全处理，尤其是在加热时。

用于直接处理热物品的坩埚钳，设计有独特的末端用于抓握，例如坩埚和热玻璃器皿。烧杯钳配有防滑橡胶握把，更适合处理较大的热烧杯。

实验室内的加热通常使用本生灯完成，能够产生可达约 700 ℃ 的火焰。为了支撑和加热玻璃器皿，使用配有铁环或铁丝网的铁环架。也可使用黏土三角安全地将坩埚放置在本生灯上。

用于转移固体粉末或晶体的药匙是一种混合了勺子和铲子的工具。可以在表面玻璃上观察物质，尽管由于其脆弱性，它并不适合加热。用于蒸发液体的蒸发皿，由耐用的瓷制成。

过滤过程使用漏斗，将混合物通过滤纸倒入，以分离固体和液体。在酸性或易与金属反应的溶液中搅拌，使用玻璃搅拌棒以避免污染。对于液体转移，移液管功能类似于滴管，但由于清洁困难而只能一次性使用。

实验室内的温度测量使用温度计进行，通常基于酒精以安全和防止破裂，而不使用水银温度计。

了解并熟悉实验室设备，包括它们的名称和功能，对于高效和安全的研究实践至关重要。这些知识是任何在实验室环境中工作的人要掌握的基础知识。

3.2 General and inspection equipment

In laboratory settings, a variety of devices are available for conducting specific functions. These include:

(1) Bunsen burners, which are utilized for heating substances.

(2) Hot plates offer more precision than Bunsen burners, as they frequently feature settings for specific temperatures or are equipped with temperature displays.

(3) Magnetic stirrers, designed for stirring substances within flasks, operate by causing a magnetic stir bar to vibrate and spin, thus mixing the solution. While often integrated into hot plates, magnetic stirrers can also function as standalone equipment.

(4) Microscopes are instrumental in magnifying objects for detailed observation.

(5) Wash bottles, typically containing distilled water—pure water devoid of minerals or impurities—serve to dispense liquids, primarily for cleaning purposes.

(6) Watch glasses are glass plates utilized for holding substances and are commonly heated on hot plates.

Organic chemistry focuses on the study of carbon-based compounds, with organic chemists concentrating on the structure, properties, and reactions of these compounds.

Essential equipment in this field includes:

(1) Photoelectron spectrometers, measuring electron energies emitted by substances.

(2) IR spectrometers, gauging the vibration and rotation of covalent bonds.

(3) Proton nuclear magnetic resonance, crucial for determining the structure of organic molecules.

Although not every organic chemistry laboratory is equipped with these specific devices, they are broadly utilized across the field.

Analytical chemistry laboratory equipment is pivotal for quantifying and understanding the properties of matter. This branch of chemistry relies extensively on analytical machinery, thus necessitating specialized laboratory equipment. Key instruments include:

(1) Gas chromatographers, separating and analyzing vaporized samples.

(2) High-powered liquid chromatographers, separating and analyzing components in mixtures.

(3) Mass spectrometers, measuring the mass-to-charge ratio of molecules/elements in samples.

Analytical chemistry is fundamentally concerned with measurements, leading to a diverse array of equipment options. The aforementioned equipment may also find applications in analytical chemistry, reflecting the interdisciplinary nature of laboratory practices.

Chemistry laboratory equipment-summary: Glassware, encompassing various measuring tools made of glass, is utilized not only for measurement but also to contain different reactants/products. This category includes Erlenmeyer flasks, beakers, volumetric flasks, graduated cylinders, graduated pipettes, burettes, test tubes, and funnels. Alongside these, other significant equipment comprises ring stands, burette clamps, utility clamps, crucible tongs, wash bottles, and watch glasses, all of which support a range of laboratory functions.

New words and expressions

hot plates	加热板	microscopes	显微镜
magnetic stirrers	磁力搅拌器	wash bottles	洗瓶

watch glasses　观察玻璃
photoelectron spectrometers　光电子能谱仪
IR spectrometers　红外光谱仪
proton nuclear magnetic resonance　质子核磁共振
gas chromatographers　气相色谱仪
high-powered liquid chromatographers　高效液相色谱仪
mass spectrometers　质谱仪
Erlenmeyer flasks　锥形瓶
beakers　烧杯
volumetric flasks　容量瓶
graduated cylinders　量筒
graduated pipettes　刻度移液管
burettes　滴定管
test tubes　试管
funnels　漏斗
ring stands　环形支架
burette clamps　滴定夹
utility clamps　通用夹具
crucible tongs　坩埚钳

译文

在实验室环境中，有多种设备可用于执行特定功能。这些包括：

（1）本生灯，用于加热物质。

（2）加热板比本生灯更精确，因为它们通常可以设置为特定温度或配有温度显示。

（3）磁力搅拌器，设计用于在烧瓶内搅拌物质，通过使磁搅拌棒振动和旋转，混合烧瓶中的溶液。虽然通常集成到加热板中，但磁力搅拌器也可以作为独立设备存在。

（4）显微镜在详细观察中放大物体发挥作用。

（5）洗瓶，通常含有蒸馏水——无矿物质或杂质的纯净水——用于分配液体，主要用于清洁。

（6）观察玻璃是用于放置物质的玻璃板，通常放在加热板上加热。

有机化学专注于碳基化合物的研究，有机化学家关注这些化合物的结构、性质和反应。该领域的关键设备包括：

（1）光电子能谱仪，测量来自不同物质的电子发射能量。

（2）红外光谱仪，测量共价键的振动和旋转。

（3）质子核磁共振，对确定有机分子的结构至关重要。

尽管并非每个有机化学实验室都配备这些特定设备，但它们在该领域广泛使用。分析化学实验室设备对于量化和理解物质的性质至关重要。这一化学分支广泛依赖于分析机械，因此需要专门的实验室设备。关键仪器包括：

（1）气相色谱仪，用于分离和分析已被汽化的样品。

（2）高效液相色谱仪，用于分离和分析混合物中的组分。

（3）质谱仪，测量样品中分子/元素的质量电荷比。

分析化学本质上关注于测量，导致设备选择多样化。上述设备也可能在分析化学中找到应用，反映了实验室实践的跨学科性质。

化学实验室设备总结：玻璃器皿，包括各种玻璃制成的测量工具，不仅用于测量，也用于容纳不同的反应物/产物。这一类别包括锥形瓶、烧杯、容量瓶、量筒、刻度移液管、滴定管、试管和漏斗。此外，其他重要设备包括环形支架、滴定夹、通用夹具、坩埚钳、洗瓶和观察玻璃，这些设备支持一系列实验室功能。

3.3　Lab safety and maintenance

3.3.1　Lab safety

Ensuring safety within the laboratory environment is paramount. Laboratories contain various potentially hazardous chemicals, necessitating strict adherence to safety protocols. Here's an expanded overview of the essential personal protective equipment (PPE) designed to safeguard individuals within laboratory settings：

（1）Eye protection：Safety goggles or protective glasses are crucial for shielding eyes from chemical exposure. Even for individuals who wear prescription glasses, additional eye protection meeting higher safety standards is required.

（2）Lab coats：These garments offer protection for both clothing and skin against harmful chemicals. Although not all laboratories mandate the use of lab coats, opting for long-sleeved shirts and pants to fully cover the skin is advisable.

（3）Gloves：Wearing protective gloves is essential for guarding hands against chemical irritants.

In the event of an incident, laboratories are equipped with emergency apparatus, whose locations may vary. It is imperative to familiarize oneself with the placement of these tools prior to commencing any experimental activities：

(1) Fire extinguishers: for extinguishing fires.

(2) Eyewash stations: for rinsing eyes of harmful chemicals.

(3) Safety showers: for removing irritants from skin and clothing.

Specialized laboratories will house additional, specific equipment. For a comprehensive understanding of laboratory safety equipment and practices, one should consult the Laboratory Safety Guide, which encompasses the following seven sections:

(1) Planning work:

1) Assess potential hazards before any experiment, identifying worst-case scenarios and the necessary precautions.

2) Understand the risks of materials used by reviewing their corrosivity, flammability, reactivity, and toxicity. Employ a project hazard review checklist for this purpose.

3) Consult Material Safety Data Sheets (MSDS) for chemical information and ensure availability of appropriate PPE.

4) Post hazard warnings and emergency contacts on the laboratory door.

5) Inspect equipment for any signs of wear or damage.

(2) Adhering to safety procedures:

1) Utilize chemical splash goggles and, when necessary, a face shield for comprehensive protection.

2) Choose gloves based on chemical compatibility.

3) Wear proper attire, avoiding shorts, open shoes, or loose clothing.

4) Avoid solitary work with hazardous chemicals.

5) Ensure adequate ventilation when using chemicals.

6) Avoid using mouth suction for pipetting; use mechanical aids instead.

7) Handle sharps carefully and dispose of them correctly.

8) Secure gas cylinders to prevent accidents.

(3) Emergency preparedness:

1) Familiarize oneself with emergency eyewash and shower locations.

2) Know multiple exit routes from the lab.

(4) Maintaining cleanliness and personal hygiene:

1) Prevent direct contact with chemicals.

2) Wash hands thoroughly after handling chemicals and before exiting the lab.

3) Keep work areas tidy and unobstructed.

(5) Safe chemical transport:

1) Utilize secondary containment for chemicals.

2) Securely transport containers.

(6) Managing unattended operations: Ensure spill or failure containment measures are in place.

(7) Reporting unsafe conditions:

1) Report all accidents and unauthorized activities.

2) Maintain vigilance for suspicious behavior.

These guidelines underscore the importance of safety, preparedness, and responsibility in maintaining a secure laboratory environment.

New words and expressions

personal protective equipment (PPE)　个人防护装备
eye protection　眼部保护
safety goggles　安全护目镜
protective glasses　防护眼镜
lab coats　实验室大衣
gloves　手套
fire extinguishers　灭火器
eyewash stations　洗眼站
safety showers　安全淋浴
Material Safety Data Sheets (MSDS)　物质安全数据表
chemical splash goggles　防化学飞溅护目镜
face shield　面罩
gas cylinders　气瓶

译文

确保实验室环境的安全至关重要。实验室包含多种潜在危险的化学品，因此严格遵守安全协议是必须的。以下是为了保护实验室人员而设计的必要个人防护装备（PPE）的扩展概述：

（1）眼部保护：安全护目镜或防护眼镜对于保护眼睛免受化学品暴露至关重要。即使是佩戴处方眼镜的个体，也需要符合更高安全标准的额外眼部保护。

（2）实验室大衣：这些服装为衣物和皮肤提供了对有害化学品的保护。虽然并非所有实验室都要求使用实验室大衣，但建议选择长袖衬衫和长裤以完全覆盖皮肤。

（3）手套：佩戴防护手套对于保护手部免受化学刺激物至关重要。

在发生事故的情况下，实验室配备了应急设备，其位置可能会有所不同。在开始任何实验活动之前，熟悉这些工具的放置位置是必须的：

（1）灭火器：用于扑灭火灾。

（2）洗眼站：用于清洗眼睛中的有害化学品。

（3）安全淋浴：用于清除皮肤和衣物上的刺激物。

专业化的实验室将配备额外的特定设备。为了全面了解实验室安全设备和实践，应该咨询实验室安全指南，该指南包含以下七个部分：

（1）规划工作：

1）在进行任何实验之前评估潜在危害，识别最坏情况及必要的预防措施。

2）通过审查其腐蚀性、易燃性、反应性和毒性，了解所使用材料的风险。为此目的使用项目危险审查清单。

3）咨询化学品的物质安全数据表（MSDS）以获取化学信息，并确保适当的 PPE 可用。

4）在实验室门上张贴危险警告和紧急联系方式。

5）检查设备是否有磨损或损坏的迹象。

（2）遵守安全程序：

1）使用化学飞溅护目镜，必要时配合面罩使用，以获得全面的保护。

2）根据化学品的兼容性选择合适的手套。

3）穿着适当的服装，避免穿短裤、敞口鞋或宽松衣物。

4）避免单独处理危险化学品。

5）使用化学品时确保足够的通风。

6）避免使用口吸来移液；改用机械辅助工具。

7）小心处理锐利物品，并正确处置。

8）固定气瓶以防止事故发生。

（3）紧急准备：

1）熟悉紧急洗眼站和淋浴的位置。

2）了解实验室的多条出口路线。

（4）保持清洁和个人卫生：

1）避免直接接触化学品。

2）在处理化学品后和离开实验室前彻底洗手。

3）保持工作区域整洁和畅通无阻。

（5）安全运输化学品：

1）对化学品使用二次容纳措施。

2）安全运输容器。

（6）管理无人看守操作：确保有溢出或故障的防护措施。

（7）报告不安全条件：

1）报告所有事故和未授权的活动。

2）对可疑行为保持警惕。

这些指南强调了在维护安全实验室环境中的安全、准备和责任的重要性。

3.3.2 Lab maintenance

Comprehensive cleaning regimen

Maintaining an immaculate laboratory isn't just straightforward and cost-effective—it's crucial, yet often underestimated in its importance. Regular cleaning routines are fundamental for the lab's overall condition and to ensure experiments are not compromised. Recommended practices include：

（1）Daily surface wipes of all equipment to remove superficial dust and spills.

（2）A thorough cleaning of equipment every week to prevent buildup of contaminants.

（3）Utilizing a 70∶30 ether-alcohol mix for meticulous cleansing of microscopes, ensuring optimal performance and accuracy.

（4）For specialized equipment, like hematology analyzers, adhere to manufacturer or lab manager guidelines for cleaning and maintenance, including periodic checks and routine cleaning.

（5）Consideration of professional services for intricate equipment cleaning to maintain functionality without the overhead of specialized training.

Adhering to these cleaning protocols ensures operational efficiency and longevity of lab equipment.

Precision in calibration

Neglecting equipment calibration compromises data integrity and can disrupt entire research projects. Regular calibration is also a safety measure in labs handling hazardous substances. Thus, it's essential to:

(1) Conduct an equipment audit to determine the necessary calibration approach, balancing between preventive maintenance and precision verification.

(2) Establish a routine calibration schedule to maintain laboratory operations at peak performance and safety.

Strategic repairs

Rather than discarding malfunctioning lab equipment, evaluate the potential for repairs or part replacements. This approach not only extends the equipment's life but is also cost-effective. Focus on items that frequently wear down, such as centrifuge rotors, filter systems, and microscope lenses, to preemptively address wear and tear.

Refurbishment as a revival strategy

When equipment functions suboptimally but isn't completely defective, refurbishing can rejuvenate its performance. The refurbishment process involves:

(1) Disassembling the equipment to clean every component thoroughly.

(2) Polishing and lubricating parts as necessary.

(3) Replacing any components showing significant wear before reassembly.

Knowledge of the equipment is vital to execute these steps effectively, enhancing the device's functionality and lifespan.

Investment in quality lab equipment

Despite rigorous maintenance, there will be instances where equipment replacement is unavoidable. Opting for cheaper alternatives might seem economical initially but often leads to increased long-term costs due to inferior durability. High-quality lab equipment, conversely, offers:

(1) Easier access to replacement parts.

(2) Simpler cleaning and refurbishment processes.

(3) A worthwhile option to consider is renting high-quality equipment for short-term needs or budget constraints, avoiding the pitfalls of cheaper but less reliable alternatives.

Implementing these advanced maintenance and procurement strategies ensures the lab remains a productive, safe, and cost-efficient environment.

New words and expressions

ether　醚
alcohol　醇
hematology analyzers　血液学分析仪
centrifuge rotors　离心机转子
filter systems　过滤系统
microscope lenses　显微镜镜片
calibration　校准
malfunctioning　故障的
lubricate　润滑
refurbishment　翻新

译文

全面清洁规程

保持实验室的洁净不仅简单、经济，而且至关重要，但这一点经常被低估。定期的清洁程序对于保持实验室整体状况和确保实验不受影响至关重要。推荐的做法包括：

（1）每天清洁所有设备的表面，以去除表面灰尘和溢出物。

（2）每周彻底清洁设备，防止污染物积累。

（3）使用70∶30的醚醇混合液仔细清洁显微镜，确保最佳性能和准确性。

（4）对于特殊设备，如血液分析仪，遵循制造商或实验室经理的清洁和维护指南，包括定期检查和例行清洁。

（5）考虑对复杂设备使用专业服务进行清洁，以维持功能性，无须专门培训。

遵循这些清洁协议确保了实验室设备的操作效率和寿命。

精确校准

忽视设备校准会损害数据完整性，并可能中断整个研究项目。定期校准也是处理危险物质的实验室的安全措施。因此，必须：

（1）进行设备审核，以确定适合每项设备的校准方法，从基本的预防性维护到更高级的精确性验证之间取得平衡。

（2）建立常规校准计划，以保持实验室在最佳性能和安全性。

策略性修复

不要急于丢弃故障设备，先评估是否可以通过更换部件或修理来解决问题。这种方法不仅可以延长设备的使用寿命，而且成本效益高。重点关注经常磨损的项目，如离心机转子、过滤系统和显微镜镜片，以预防磨损。

翻新作为复活策略

当设备功能下降但并非完全损坏时，翻新可以恢复其性能。翻新过程包括：
（1）拆卸设备以彻底清洁每个组件。
（2）必要时进行打磨和润滑。
（3）在重新组装前更换任何显示明显磨损的部件。
了解设备至关重要，以有效执行这些步骤，提高设备的功能性和寿命。

投资高质量实验室设备

尽管进行了严格的维护，但有时仍需更换设备。选择看似经济的便宜选项可能初看起来省钱，但经常会导致更高的长期成本，因为这些便宜的设备耐用性差。相反，选择高质量的实验室设备通常提供：
（1）更容易找到替换部件。
（2）清洁和翻新过程更简单。
（3）如果您只打算短期使用设备或预算非常紧张，考虑租用高质量设备而不是购买便宜的替代品可能是一个明智的选择。
实施这些先进的维护和采购策略确保实验室保持高效、安全和高成本效益的环境。

3.4 How to write a research paper?

This guide embarks on an exploration of essay writing, transitioning from a broad overview to a nuanced discussion of the various essay types that students might encounter throughout their academic journey. Within the realms of academia, four primary essay genres stand out: descriptive, narrative, expository, and argumentative essays. These forms, also known as modes of discourse, serve as common assignments in writing courses. Despite some academic debate over their use, the widespread acceptance and necessity for students to master these formats are recognized and supported by educational resources like the Purdue OWL.

Comprehensive guide to essay writing

Essays represent a fundamental aspect of academic assignments, a challenge and opportunity that every student faces. Mastering essay writing early in one's educational career is not only wise but essential for academic success.

Essays serve as a medium for students to engage in rigorous intellectual exercises, enabling the cultivation of various skills including critical reading, analytical thinking, comparison and contrast, persuasion, brevity, clarity, and exposition. Through this engagement, students are positioned to excel in crafting insightful and coherent pieces.

The essence of an essay is to foster the development of ideas and concepts with minimal external guidance, offering a counterpoint to the structure of research papers. Essays demand conciseness and a clear objective, leaving no space for meandering thoughts but instead promoting focused and engaging writing.

This resource aims to equip students with the knowledge and confidence needed to navigate the landscape of essay writing by introducing common essay types. This includes an overview of:

(1) Expository essays: Essays that explain or analyze a topic with a focus on clarity and information dissemination.

(2) Descriptive essays: These essays paint a picture through detailed observations and descriptions, evoking senses to present a subject.

(3) Narrative essays: Telling stories with a clear purpose, narrative essays transport readers through the author's personal experiences or imagined scenarios.

(4) Argumentative (persuasive) essays: These essays take a stance on a topic and argue in favor of it, aiming to persuade the reader through evidence and reasoning.

Through delving into these genres, this guide illuminates the diverse spectrum of essay writing, encouraging students to experiment and refine their writing skills across different contexts and purposes.

New words and expressions

descriptive 描述性的
narrative 叙述性的
expository 说明性的
argumentative 论证性的
modes of discourse 话语模式
close reading 精读
analysis 分析
comparison and contrast 对比和比较
persuasion 说服
conciseness 简洁
clarity 清晰
exposition 阐述

译文

这份指南开始探讨论文写作，从广泛概述过渡到对学生在整个学术旅程中可能遇到的各种论文类型的深入讨论。在学术领域内，4 种主要的论文体裁脱颖而出：描述性论文、叙述性论文、说明性论文和论证性论文。这些形式，也称为话语模式，是写作课程中的常见任务。尽管对它们的使用存在一些学术争议，但学生掌握这些格式的广泛接受和必要性得到了 Purdue OWL 等教育资源的认可和支持。

论文写作全面指南

论文是学术任务中的基本组成部分，是每位学生面临的挑战和机会。在个人的教育生涯早期掌握论文写作不仅是明智的，而且对学术成功至关重要。

论文作为一种媒介，使学生参与到严谨的智力练习中，使他们能够培养包括批判性阅读、分析思维、对比和比较、说服、简洁、清晰和阐述等多种技能。通过这种参与，学生将能够在制作有洞察力和连贯的作品方面表现出色。

论文的本质是在最少的外部指导下促进思想和概念的发展，提供与研究论文结构相反的视角。论文要求简洁明了，目标明确，没有空间让学生的思维漫游或偏离其目的；写作必须是刻意和有趣的。

这个资源旨在通过介绍常见的论文类型，使学生熟悉并自信地掌握论文写作的过程。这包括对以下类型的概述：

（1）说明性论文：解释或分析一个主题的论文，注重清晰和信息传播。

（2）描述性论文：通过详细的观察和描述绘制画面的论文，唤起感官以呈现一个主题。

（3）叙述性论文：以清晰的目的讲述故事，叙述性论文通过作者的个人经历或想象场景将读者带入其中。

（4）论证性（说服性）论文：这些论文对一个话题采取立场并为之辩护，旨在通过证据和推理说服读者。

通过深入这些体裁，本指南阐明了论文写作的多样性，鼓励学生在不同的语境和目的中实验和完善他们的写作技巧。

3.4.1　What is an expository essay?

The expository essay, a pivotal genre within academic writing, necessitates that students delve into an idea, scrutinize evidence, elaborate on the concept, and present an argument related to that idea in a manner that's both lucid and succinct. Achieving this requires employing various strategies, including but not limited to, comparison and contrast, providing definitions, citing examples, and conducting cause and effect analysis.

It's worth mentioning that this type of essay is frequently utilized as a mechanism for classroom assessment and commonly appears in several examination contexts.

Foundational structure of the expository essay:

（1）Introductory thesis statement: The essay is anchored by a thesis statement in its opening paragraph that is both precise and defined. It's imperative that the thesis is narrowly focused to adhere to the assignment's stipulations. A well-crafted thesis is critical; without it, crafting an effective or persuasive essay becomes significantly challenging.

（2）Coherent transitions: Smooth and logical transitions that connect the introduction, body, and conclusion are essential. These transitions act as the glue that maintains the essay's structure, ensuring a seamless flow of ideas. Without these, the reader might lose track of the argument, leading to a disjointed reading experience.

（3）Body paragraphs with supportive evidence: Each paragraph should concentrate on a single main idea, backed by appropriate evidence, to maintain clarity and direction. This approach not only enhances the essay's readability but also ensures that it remains focused and impactful. Importantly, every paragraph should logically relate back to the thesis introduced at the beginning.

(4) Varied forms of evidence: Supportive evidence, be it factual, logical, statistical, or anecdotal, is crucial. While expository essays might not always allow for extensive statistical or factual data due to limited preparation time, incorporating a diverse range of evidence strengthens the argument.

(5) Creative engagement: Despite the structured nature of expository essays, incorporating elements of creativity and artfulness can elevate the writing. Avoid becoming overly fixated on the essay's formula at the cost of engaging content. Strive to make a memorable impression on your readers with your creativity and insight.

(6) Insightful conclusion: The conclusion should extend beyond merely reiterating the thesis; it should reinterpret it in the context of the evidence discussed. This segment is vital as it leaves a lasting impression on the reader. It must be cogent and conclusive, without introducing new information, synthesizing the essay's content to draw a compelling conclusion.

While the five-paragraph format is a common and straightforward method for structuring expository essays, comprising an introduction, three body paragraphs with evidence, and a conclusion, it's not the sole approach. Flexibility in structure, while maintaining the essay's core objectives, can contribute to a more nuanced and persuasive argument.

New words and expressions

expository essay　说明性论文
evidence　证据
comparison and contrast　比较与对比
definition　定义
example　例证
cause and effect analysis　因果分析
thesis statement　论文陈述
transitions　过渡
body paragraphs　主体段落

evidential support　证据支持
factual　事实的
logical　逻辑的
statistical　统计的
anecdotal　轶事的
creativity　创造性
artfulness　巧妙
conclusion　结论

译文

说明性论文是学术写作中的一个关键体裁，要求学生探讨一个观点、评估证据、详细阐述这个概念，并以清晰简洁的方式就这个观点提出论证。要做到这一点，可以采用多种策略，包括比较与对比、提供定义、列举例证、进行因果分析等。

值得一提的是，这种类型的论文通常被用作课堂评估的工具，并常出现在多种考试格式中。

说明性论文的基本结构包括：

（1）开篇论文陈述：文章由第一段的清晰、简洁且明确的论文陈述作为锚点。论文陈述必须适当地缩小范围以符合作业的指导方针。如果学生在这部分没有掌握好，撰写一个有效或有说服力的论文将会非常困难。

（2）连贯的过渡：连接引言、主体和结论的清晰逻辑过渡至关重要。这些过渡像胶水一样把论文的结构固定住，确保思想的流畅过渡。没有逻辑的进展，读者无法跟随文章的论证，结构就会崩溃。

（3）包含证据支持的主体段落：每个段落应限于对一个总体观点的阐述。这将确保整篇论文的方向清晰。更重要的是，这种简洁性为读者创造了易读性。需要注意的是，文章主体的每个段落都必须与开头段落中的论文陈述有逻辑上的联系。

（4）多样化的证据形式：无论是事实的、逻辑的、统计的还是轶事的，证据都是至关重要的。由于说明性论文可能因为准备时间有限而不允许大量的统计或事实证据，因此，采用多样化的证据可以增强论点。

（5）创造性的运用：尽管说明性论文的结构性质，融入创造性和技巧可以提升写作水平。避免过分固守于论文的公式性，而牺牲了撰写有趣内容的可能。以你的创造性和洞察力努力在评估你论文的人们心中留下持久印象。

（6）富有洞察力的结论：结论应超出仅仅重申论文陈述，而是在所讨论的证据的背景下重新解读它。这一部分至关重要，因为它将在读者心中留下最直接的印象。因此，它必须是有效且逻辑的。结论中不要引入任何新信息，而是综合文章主体中的信息，得出有说服力的结论。

虽然五段论是一种常见且简单的说明性论文结构方法，包括一个引言段、三个带有证据的主体段落和一个结论，但这并非写作这类论文的唯一公式。在保持论文核心目标的同时，灵活地调整结构可以为论证贡献更多的细腻和说服力。

3.4.2 What is a descriptive essay?

The descriptive essay stands as a vibrant canvas in the realm of academic writing, inviting students to articulate the essence of objects, individuals, locales, experiences, emotions, or scenarios through their prose. This particular genre not only nurtures the student's capacity to forge a narrative around a distinct experience but also grants substantial creative liberty aimed at crafting an image so vivid and evocative it resonates within the reader's imagination.

Guidelines for crafting a descriptive essay:

(1) Brainstorming session: Before diving into the descriptive endeavor, it's crucial to gather your thoughts. For example, if tasked with depicting your favorite food, jot down preliminary ideas that capture its essence. Selecting pizza could lead to noting attributes like sauce, cheese, crust, and toppings. These initial thoughts lay the groundwork for a more structured and detailed description.

(2) Precision in language: Opt for language that's both precise and evocative. The selection of words should be deliberate, aiming to convey your subject with clarity and depth.

(3) Vivid language use: Elevate your description with language that paints a more dynamic picture. Opt for "stallion" over "horse", "tempestuous" rather than "violent", or "miserly" instead of "cheap". Such choices enrich the imagery and nuance, enhancing the reader's engagement.

(4) Engage the senses: To truly bring your description to life, appeal to the reader's senses. Detail not just how your subject looks, but also its scent, texture, sound, and taste. This sensory engagement deepens the reader's connection to your narrative.

(5) Emotional connectivity: Delve into emotions and feelings associated with your topic. By tapping into universal experiences of joy, loss, or contentment, you forge a deeper connection with your audience, elevating your descriptive power.

(6) Memorable impressions: Aim to leave the reader with a lasting impression, one that not only evokes a sense of familiarity but also a profound appreciation. If, after reading your essay, the reader craves the pizza you've described, you've mastered the art of descriptive writing.

(7) Structured coherence: While it's tempting to meander through various sensations and emotions, maintaining a coherent and organized structure is paramount. This ensures the reader emerges with a clear understanding of your subject, fully appreciating the depth of your description.

By adhering to these guidelines, you not only enhance the descriptive quality of your essays but also enrich the reader's experience, making your writing not just informative but truly immersive.

New words and expressions

descriptive essay 描述性论文
brainstorming 头脑风暴
precision in language 语言的精确性
vivid language 生动的语言
engage the senses 吸引感官
emotional connectivity 情感联系
structured coherence 结构性的连贯性
narrative 叙述
creative liberty 创造性自由
imagination 想象力
attributes 属性
sensory engagement 感官参与
universal experiences 普遍经验

译文

描述性论文在学术写作领域中如同一块生动的画布,它邀请学生通过文字表达物体、人物、地点、经历、情感或情境的本质。这一特定体裁不仅培养了学生围绕一个独特体验构建叙述的能力,而且提供了大量的创造性自由,旨在打造出如此生动和感人的图像,使之在读者的想象中产生共鸣。

撰写描述性论文的指南:

(1) 头脑风暴:在深入描述之前,收集你的思维至关重要。例如,如果任务是描述你最喜欢的食物,请先记下捕捉其精髓的初步想法。选择比萨饼可能会引出如酱料、奶酪、面皮和配料等属性的记录。这些初步的思考为更有结构和详细的描述奠定了基础。

(2) 精确的语言:选择既精确又生动的语言。单词的选择应该是经过深思熟虑的,旨在以清晰和深度传达你打算描述的内容。

(3) 使用生动的语言:通过使用更动态的语言来提升你的描述。选择"骏马"而不是"马","狂暴"而不是"暴力",或者"吝啬"而不是"便宜"。这样的选择丰富了图像和细微含义,增强了读者的参与度。

(4)吸引感官:要真正让你的描述栩栩如生,需要吸引读者的感官。不仅描述物体的外观,还要描述它的气味、质感、声音和味道,用感官美化这一刻。

(5)情感联系:深入探索与你的主题相关的情感或感受。通过触及快乐、失落或满足等普遍经验,你与观众建立了更深层次的联系,提升了你的描述力。

(6)留下难忘的印象:你的目标之一是在读者中唤起强烈的熟悉感和赞赏感。如果你的读者在阅读完论文后渴望你刚刚描述的比萨饼,那么你就成功地掌握了描述性写作的艺术。

(7)有组织的连贯性:虽然很容易陷入对情感和感官的杂乱无章的描述,但保持连贯有序的结构是至关重要的。这确保读者能够清晰地理解你试图描述的内容,充分欣赏你描述的深度。

遵循这些指南,不仅可以提升你论文的描述质量,还可以丰富读者的体验,使你的写作不仅仅是提供信息,而是真正沉浸式的。

3.4.3　What is a narrative essay?

Crafting a narrative essay is akin to weaving a story from the fabric of personal experience. These essays, rich in anecdotes and personal revelations, provide a platform for students to unfold their creativity and often convey profound messages in remarkably touching ways.

Essentials for crafting a narrative essay:

(1) Storytelling framework: If adopting a storytelling approach, your essay should encompass all traditional elements of a narrative. This includes an enticing introduction, a compelling plot, vivid characters, a descriptive setting, a pivotal climax, and a conclusive resolution.

(2) Exceptions to the story form: While narrative essays typically mirror the structure of stories, there are instances, such as when writing a book report, where this format does not apply. In such cases, the focus shifts towards delivering an informative narrative rather than adhering strictly to the storytelling format.

(3) Purposeful writing: Every narrative should serve a purpose or convey a message. Consider this the thesis of your narrative. Without a meaningful core, the narrative may lose its significance. Ask yourself, what is the deeper message or insight you wish to

share through your story?

(4) Distinct perspective: Narrative essays commonly reflect the author's viewpoint. However, the beauty of narrative writing lies in the possibility of exploring different perspectives. Let your creativity shine by experimenting with various narrative voices, thereby enriching your story's depth.

(5) Clarity and precision in language: The effectiveness of a narrative essay is significantly enhanced by the deliberate choice of language. Aim for precision and artistry in your word choice to stir specific emotions and invoke vivid imagery, much like in descriptive writing.

(6) Personal voice: The use of first-person narration ("I") is encouraged to add authenticity and intimacy to your narrative. However, this technique should be used judiciously to maintain clarity and impact without compromising on the narrative's articulation.

(7) Structured organization: Begin with a clear introduction that sets the tone and direction for the essay. Avoid ambiguity about your narrative's purpose. As the narrator, you have the power to steer your narrative, ensuring that it unfolds in a manner that aligns with your intended message.

By adhering to these guidelines, you can transform personal experiences and imaginative ideas into compelling narrative essays that captivate and resonate with readers, guiding them through a journey of discovery, reflection, and emotional connection.

New words and expressions

narrative essay　叙述性论文
anecdotal　轶事的
experiential　经验的
creativity　创造性
storytelling framework　讲故事的框架
introduction　引言
plot　情节
characters　人物
setting　背景设定

climax　高潮
conclusion　结论
informative narrative　信息性叙述
purposeful writing　有目的的写作
distinct perspective　独特视角
clarity and precision in language　语言的清晰与精确
personal voice　个人声音
structured organization　有结构的组织

译文

编写叙述性论文就像从个人经验的织物中编织一个故事。这些论文充满了轶事和个人启示,为学生提供了一个展开创造力的平台,并且通常以非常感人的方式传达深刻的信息。

编写叙述性论文的要点:

(1) 讲故事的框架:如果采用讲故事的方式,你的论文应该包含所有传统的叙事元素。这包括一个吸引人的引言、引人入胜的情节、生动的人物、详细的背景设定、关键的高潮和结局。

(2) 故事形式的例外:虽然叙述性论文通常反映故事的结构,但在某些情况下,例如撰写书评时,这种格式并不适用。在这种情况下,重点转向提供一个信息性叙述,而不是严格遵循讲故事的格式。

(3) 有目的的写作:每个叙述都应该有一个目的或传达一个信息。将这视为你叙述的论点。如果你所叙述的没有深层次的意义,那么为什么要叙述它呢?

(4) 独特视角:叙述性论文通常反映作者的观点。然而,叙述写作的美在于探索不同的视角的可能性。通过尝试各种叙述声音,让你的创造力闪耀,从而增加你的故事的深度。

(5) 语言的清晰与精确:通过精心挑选语言,叙述性论文的效果显著增强。目标是通过精确和艺术性的词语选择,激发特定的情感并唤起生动的图像,这与描述性写作非常相似。

(6) 个人声音:鼓励使用第一人称叙述("我")以增加你的叙述的真实性和亲密感。然而,这种技巧应该谨慎使用,以保持清晰和影响力,而不会损害叙述的表达。

(7) 有结构的组织:从一个清晰的引言开始,为论文的其余部分设定基调。不要让读者对你的叙述目的感到猜测。记住,你控制着论文的方向,所以按照你的愿望引导它。

通过遵循这些指南,你可以将个人经历和想象中的想法转化为吸引人的叙述性论文,吸引并与读者产生共鸣,引导他们进行发现、反思和情感连接的旅程。

3.4.4　What is an argumentative essay?

The genre of the argumentative essay is a profound exploration of writing that mandates students to delve into a particular topic; to gather, create, and critically assess evidence; and to concisely articulate a stance on the matter at hand. It is essential to recognize the nuanced difference between the argumentative essay and its close relative, the expository essay. Despite their similarities, these two forms diverge significantly in the scope of preparatory work and the depth of research required. The argumentative essay, often a hallmark of capstone or advanced writing courses, demands extensive and thorough research. In contrast, expository essays, which are briefer and require less research, are typically deployed for in-class assignments or standardized tests like the GED or GRE.

Embarking on an argumentative essay involves a deep foray into both literature and previously published materials. Such assignments might also engage students in empirical research, including interviews, surveys, observations, or experimental methods, facilitating a comprehensive understanding of the topic from multiple perspectives. This broad-based research empowers students to form a well-supported stance. An argumentative essay's success hinges on its foundation: A well-defined thesis supported by rigorous reasoning, regardless of the research's breadth or type.

The architecture of an argumentative essay is meticulously constructed, beginning with a precise, succinct thesis statement in the opening paragraph. This introductory section should lay out the context, underscore the topic's significance, and culminate in the thesis statement, which must be sufficiently narrowed to meet the assignment's criteria. Mastery of this segment is critical for crafting an impactful and convincing essay.

The essay's coherence is maintained through clear, logical transitions that seamlessly connect the introduction, body, and conclusion. These transitions are pivotal, acting as the cohesive force that underpins the argument's flow and structural integrity. The body of the essay, organized into paragraphs each focused on a singular idea, provides evidential support that directly ties back to the thesis, ensuring clarity and ease of understanding for the audience. Moreover, a comprehensive argumentative essay will venture beyond merely presenting evidence in support of the thesis to also consider and articulate counterpoints and differing views on the subject matter.

The argumentative essay demands a rich tapestry of evidence—factual, logical, statistical, or anecdotal—to bolster the thesis while engaging with a spectrum of viewpoints. Ethically, it's imperative to acknowledge evidence that may challenge the thesis, rather than omitting it, thereby enriching the argument's depth and credibility.

The essay culminates in a conclusion that transcends mere repetition of the thesis, instead reevaluating it in the context of the evidence presented. This critical juncture seeks to leave a lasting impression, emphasizing the topic's relevance, summarizing key arguments, and possibly suggesting directions for future research.

While the five-paragraph essay offers a straightforward framework for constructing an argumentative piece, it is by no means exhaustive, especially for essays tackling complex issues or extensive research. Longer argumentative essays will naturally require a more elaborate structure, possibly encompassing a broader discussion on the topic's background, the credibility of sources, and various viewpoints before drawing to a close. Such essays underscore the dynamic nature of argumentative writing, accommodating a range of complexities and depths to foster critical thinking and articulate debate.

New words and expressions

argumentative essay　论证性作文
expository essay　说明性作文
capstone　顶石项目
advanced composition courses　高级写作课程
empirical research　实证研究
thesis statement　论文陈述/主题句
transitions　过渡
evidential support　证据支持
logical progression　逻辑进展
counterpoints　反论点

译文

　　论证性作文是一种写作类型，要求学生探索一个主题；收集、产生和评估证据；并以简明的方式确立对该主题的立场。需要注意的是，论证性作文和说明性作文之间可能会发生一些混淆。这两种类型相似，但论证性作文在预写（创作）和研究所涉及的量上与说明性作文不同。论证性作文通常作为第一年写作或高级写作课程的顶石预式项目或最终项目，并涉及长时间、详尽的研究。说明性作文涉及的研究较少，篇幅也更短。说明性作文常用于课堂写作练习或测试，如 GED 或 GRE。

论证性作文作业通常要求对文献或之前发布的材料进行广泛的研究。这样的作业也可能使学生参与到实证研究中，包括访谈、调查、观察或实验方法，从而从多个角度全面理解主题。这种广泛的研究使学生能够形成一个有充分支持的立场。论证性作文的成功依赖于其基础：一个明确定义的论文陈述，由严密的推理支持，无论研究的广度或类型如何。

论证性作文的结构是精心构建的，从开头段落的一个准确、简洁的论文陈述开始。这个引言部分应该展示背景，强调主题的重要性，并以足够细化以满足作业标准的论文陈述结束。掌握这一部分对于编写一个有影响力和令人信服的论文至关重要。

通过清晰、逻辑的过渡在引言、主体和结论之间无缝连接，维持了作文的连贯性。这些过渡是至关重要的，作为支撑论点流程和结构完整性的凝聚力。作文的主体部分组织成每段聚焦一个单一想法的段落，提供直接与论文陈述相关的证据支持，确保了观众的清晰度和易于理解。此外，一个全面的论证性作文将超越仅仅提供支持论文的证据，还会考虑和阐述主题上的反论点和不同观点。

论证性作文要求一个丰富的证据帷幕——事实的、逻辑的、统计的或轶事的——以加强论文陈述，同时涉及一系列的观点。从伦理上讲，承认可能挑战论文陈述的证据，而不是省略它，是必要的，从而丰富了论点的深度和可信度。

作文在结论中达到高潮，不仅仅是重复论文陈述，而是在所呈现的证据的背景下重新评估它。这一关键时刻寻求留下持久印象，强调主题的相关性，总结关键论点，并可能提出未来研究的方向。

虽然五段式作文提供了一个构建论证性作品的直接框架，但它绝不是详尽无遗的，尤其是对于处理复杂问题或广泛研究的论文。较长的论证性作文自然会需要一个更复杂的结构，可能包括对主题背景、信息来源的可信度和各种观点的更广泛讨论，然后才结束。这样的论文强调了论证性写作的动态性，适应一系列的复杂性和深度，以培养批判性思维和明确的辩论。

3.5 Research methodology

In the composition of a thesis or dissertation, an essential section to be addressed is the methodology, which delineates the methods utilized in conducting the research. This section explicates the procedures and approaches employed, enabling an assessment of the study's reliability and validity by the readers. It should encapsulate the research type conducted, detailing: Firstly, the data collection methods; secondly, the data analysis techniques; thirdly, the tools or materials employed in the research process; and lastly, the rationale behind the selection of these methods. The narrative in the methodology section is typically conveyed in the past tense.

Definition of research methodology

Research methodology refers to the specific procedures or techniques adopted to identify, select, process, and analyze information about a topic within a research paper. This section grants the reader the ability to critically assess the study's overall validity and reliability.

Positioning of the methodology section in a research paper: In the structure of a scientific paper, the methodology section is positioned after the introduction and precedes the results, discussion, and conclusion sections. This arrangement is also applicable to theses, dissertations, or research proposals, where, depending on the document's length and type, a literature review or theoretical framework may precede the methodology.

General types of research methodology

Research methodologies vary, including qualitative research methodology, which is descriptive and subjective, emphasizing observation and description over factual accuracy. Its primary goal is to assess individuals' knowledge, attitudes, behaviors, and opinions concerning the research topic, utilizing methods such as grounded research, case studies, action research, disclosure analysis, and ethnography, focusing on the quality of the phenomenon observed.

Characteristics of qualitative research

In qualitative research, understanding an event within its context is paramount. Researchers immerse in the setting, embracing natural, unpredefined contexts of inquiry. It is an interactive process, aiming to capture the participants' perspectives and experiences as accurately as possible.

Strengths and limitations of qualitative research

Qualitative research offers in-depth insights into the problem, focusing on suggesting causes, effects, and possible relationships without relying on statistical data. However, it faces challenges such as lengthy data collection and analysis processes, potential bias from researcher presence, and concerns about the reliability and validity of findings.

Quantitative research methodology, on the other hand, systematically tests research hypotheses through numerical data, incorporating methods like laboratory experiments, econometric calculations, surveys, and simulations. It emphasizes measurement and quantity as pivotal aspects.

Strengths and limitations of quantitative research

Quantitative methods are known for their precision, control, and ability to generate causal statements through statistical analysis. Yet, they may overlook the human experience, losing control over variables and not accounting for individuals' unique capacities to construct meaning.

The methodology section is more than a description of data gathering and analysis; it encompasses the overarching approaches and perspectives of the research process. For those aiming to present a research proposal for studying abroad, it's crucial to clearly articulate the chosen study methods. This guide offers insights and highlights pitfalls to avoid for crafting an effective research methodology.

New words and expressions

thesis or dissertation 论文或学位论文	quantitative research methodology 定量研究方法论
research methodology 研究方法论	grounded research 基础研究
data collection methods 数据收集方法	case study 案例研究
data analysis techniques 数据分析技术	action research 行动研究
research process 研究过程	disclosure analysis 披露分析
qualitative research methodology 定性研究方法论	ethnography 民族志
	laboratory experiments 实验室实验

econometric calculations	计量经济学计算	surveys	调查研究
		simulations	模拟

译文

在撰写论文或学位论文时，必须讨论使用的研究方法。研究方法论部分阐述了采取的行动及其执行方式，允许读者评估研究的可靠性和有效性。它应包括进行的研究类型：第一，如何收集数据；第二，如何分析数据；第三，在研究中使用的任何工具或材料；第四，选择这些方法的理由。方法论部分通常应使用过去时态书写。

研究方法论的定义

研究方法论是用于识别、选择、处理和分析研究论文中某个主题的信息的具体程序或技术。这一部分允许读者批判性地评估研究的整体有效性和可靠性。

研究论文中方法论部分的位置：在科学论文中，方法论部分总是位于引言之后及结果、讨论和结论之前。这一基本结构也适用于论文、学位论文或研究提案。根据文档的长度和类型，在方法论之前还会有文献综述或理论框架。

研究方法论的一般类型

研究方法论种类繁多，包括定性研究方法论，它是描述性和主观性的，强调观察和描述超过事实的准确性。其主要目标是评估个体关于研究主题的知识、态度、行为和意见，采用诸如基础研究、案例研究、行动研究、披露分析和民族志等方法，专注于观察到的现象的质量。

定性研究的特点

在定性研究中，理解事件在其上下文中的重要性至关重要。研究人员沉浸在环境中，拥抱自然、未预定义的查询上下文。这是一个互动过程，旨在尽可能准确地捕捉参与者的观点和经历。

定性研究的优势和局限性

定性研究提供了对问题的深入洞察，关注于提出原因、效果和可能的关系，而不依赖统计数据。然而，它面临挑战，如数据收集和分析过程的漫长、研究人员存在的潜在偏见，以及对发现的可靠性和有效性的担忧。

另一方面，定量研究方法论通过数值数据系统地测试研究假设，采用诸如实验室实验、计量经济学计算、调查和模拟等方法。它强调测量和数量作为关键方面。

定量研究的优势和局限性

定量方法以其精确性、控制和通过统计分析生成因果声明的能力而闻名。然而，它们可能忽略了人类经验，失去对变量的控制，并且没有考虑个体构建意义的独特能力。

方法论部分不仅仅是数据收集和分析的描述，它涵盖了研究过程的总体方法和视角。对于那些旨在提交留学研究提案的人来说，清晰地阐述所选择的研究方法至关重要。本指南提供了见解，并强调了为制定有效的研究方法论应避免的陷阱。

Crafting an effective methodology section

(1) Introduction to methods: Present the methodological framework utilized for addressing the research question. The framework could adopt quantitative, qualitative, or mixed-method approaches, as outlined in prior sections.

(2) Methodological alignment: Clarify the relevance of the selected methodological approach to the comprehensive research design, ensuring the methodology's alignment with the research question is transparent. The methodology chosen should be well-suited to meet the objectives of the research paper and address the posed research question effectively.

(3) Instrumentation detailing: Detail the instruments designated for data collection, elucidating their application within the research. These may encompass surveys, interview questionnaires, observation techniques, etc. For methods involving archival research or data analysis from existing sources, provide context on the documentation, the original researcher, and the methodology behind the data collection.

(4) Analysis procedure: Describe the analytical process for the data collected, indicating whether statistical analysis or theoretical exploration will be employed to

interpret observed phenomena.

(5) Methodological background: For methods potentially unfamiliar to the audience, offer comprehensive background information to aid in understanding.

(6) Sampling rationale: Elucidate the reasoning behind the chosen sampling strategy, specifying why certain methods or procedures were selected for the research. This includes detailing the selection process for interview participants and the conduct of interviews.

(7) Addressing limitations: Acknowledge any potential limitations within the research, including practical challenges that may impact data collection. Discuss anticipated issues and justify the continued use of the methodology despite identified risks.

(8) Methodology section writing avoidances:

1) Omission of irrelevant details: Refrain from incorporating extraneous information that does not enhance understanding of the employed methods.

2) Precision and comprehensiveness: Ensure the methodology section is direct and exhaustive without delving into superfluous explanations of basic or well-understood procedures, unless they deviate from conventional methods and are likely unfamiliar to the audience.

3) Acknowledgement of challenges: Do not overlook obstacles encountered during data collection. Describe the approach to managing these issues, providing insight into the research process's adaptability and rigor.

New words and expressions

methodological framework　方法论框架
quantitative, qualitative, or mixed-method approaches　定量、定性或混合方法途径
instrumentation　仪器设备
data collection　数据收集
surveys　调查
interview questionnaires　访谈问卷
observation techniques　观察技术
archival research　档案研究
data analysis　数据分析
statistical analysis　统计分析
theoretical exploration　理论探索
sampling strategy　抽样策略
limitations　限制
data collection challenges　数据收集挑战

译文

（1）方法介绍：展示用于解决研究问题的方法论框架。该框架可以采取定量、定性或混合方法途径，如之前章节所概述。

（2）方法论对齐：明确所选方法论途径与整体研究设计的相关性，确保方法论与研究问题的一致性是透明的。所选择的方法论应能够有效地满足研究论文的目标并解决提出的研究问题。

（3）仪器详述：详细描述用于数据收集的仪器，并阐明其在研究中的应用。这些可能包括调查问卷、访谈问卷、观察技巧等。对于涉及档案研究或现有数据分析的方法，提供关于文档的背景、原始研究者及数据收集背后的方法论。

（4）分析过程：描述对收集数据的分析过程，指出将采用统计分析或理论探索来解释观察到的现象。

（5）方法论背景：对可能不为读者熟悉的方法提供全面的背景信息，以帮助理解。

（6）抽样理由：阐述选择特定抽样策略背后的原因，指明为什么选择某些方法或程序进行研究。这包括详细说明访谈参与者的选择过程及访谈的进行方式。

（7）讨论限制：承认研究中可能存在的任何潜在限制，包括可能影响数据收集的实际挑战。讨论预期的问题并证明尽管识别出风险仍然决定使用该方法论。

（8）方法论部分写作应避免的内容：

1）避免无关细节：避免加入不增加对所用方法理解的额外信息。

2）精确与全面：确保方法论部分直接而详尽，不深入讲解基础或广为人知的程序，除非这些程序偏离常规方法且可能对读者不熟悉。

3）承认挑战：不要忽视在数据收集过程中遇到的障碍。描述管理这些问题的方法，提供研究过程适应性和严格性的洞察。

3.6　Research integrity and ethics

Before World War Ⅱ, societal engagement with the intricacies of research and the ethics governing it was minimal. Researchers enjoyed a considerable degree of autonomy in their work, seldom questioned on their methodologies or the moral compass guiding their endeavors. However, the harrowing revelations of the Nuremberg trials, spotlighting the inhuman experiments in Nazi concentration camps, marked a turning point. These revelations awakened a collective conscience about ethical standards in research.

The aftermath saw a gradual but undeniable shift towards accountability and ethical scrutiny in research practices. Although researchers retained some discretion in adhering to ethical guidelines, the establishment of the Declaration of Helsinki in 1964 signaled a concerted international effort towards a unified ethical framework. This declaration, which has undergone several revisions, underscores the necessity of scientific review before commencing any research project.

The release of the Belmont Report in 1979 was a further watershed moment, highlighting the deficiencies in individual researchers' commitment to ethical standards. This prompted governmental intervention across the globe, leading to the creation and enforcement of regulations designed to foster responsible research conduct.

In 2010, the Singapore Statement on Research Integrity emerged from a global gathering of delegates, aiming for a harmonized approach to regulation that respects independent national laws. Research integrity, as defined in this context, transcends mere compliance with external mandates, emphasizing the intrinsic value of ethical principles and professional standards throughout the research lifecycle—from planning through to dissemination.

The landscape of responsible research is informed by a tapestry of guidelines, including professional codes, government regulations, institutional policies, and, crucially, the personal values and ethical convictions of researchers themselves. At the heart of these guidelines are fundamental ethical principles aimed at ensuring research is conducted responsibly:

(1) Honesty: Uphold truthfulness in all aspects of research, from data reporting to communications with peers and the public.

(2) Objectivity: Mitigate bias across all research phases, acknowledging personal or financial interests that could influence outcomes.

(3) Integrity: Adhere to commitments and act with sincerity, ensuring consistency between beliefs and actions.

(4) Carefulness: Exercise diligence to avoid errors, and thoroughly vet both personal work and that of colleagues.

(5) Openness: Foster a culture of transparency and receptivity to critique, sharing resources and findings openly.

(6) Intellectual property: Respect copyrights and other intellectual property rights, crediting all contributions appropriately.

(7) Confidentiality: Safeguard sensitive information, respecting the privacy and trust of all stakeholders.

(8) Responsible publication and mentoring: Publish with the intention of advancing knowledge and mentor with integrity, fostering the development of students and peers.

(9) Respect, social responsibility, and non-discrimination: Treat colleagues with respect, commit to societal welfare, and ensure fairness and equity in all dealings.

(10) Competence and legality: Continuously seek personal and professional growth while adhering to legal and regulatory standards.

(11) Care for animals and human subjects: Approach research involving animals and humans with the utmost respect and ethical consideration, prioritizing their well-being and rights.

This evolution in research ethics and integrity reflects a growing recognition of the need for a principled approach that balances scientific inquiry with the welfare of society and the natural world.

New words and expressions

World War II 第二次世界大战
Nuremberg trials 纽伦堡审判
Nazi concentration camps 纳粹集中营
Declaration of Helsinki 赫尔辛基宣言
Belmont Report 贝尔蒙特报告
Singapore Statement on Research Integrity 新加坡研究诚信声明
research integrity 研究诚信
ethical standards 伦理标准
scientific review 科学审查
intellectual property 知识产权
confidentiality 保密性
responsible publication 负责任的发布
mentoring 指导
social responsibility 社会责任
non-discrimination 非歧视
legality 合法性
human subjects 人类受试者

译文

在第二次世界大战之前，社会对研究的复杂性及其伦理规范的关注非常有限。研究者们在工作中享有相当大的自主权，他们的研究方法或道德指南很少受到质疑。然而，纽伦堡审判揭露的惊人事实——纳粹集中营中的非人实验，标志着一个转折点。这些揭露唤醒了人们对研究伦理标准的集体意识。

其后，研究实践中向着负责任和伦理审查方向的转变逐渐而明确地发生了。尽管研究者在遵循伦理指南上仍有一定的自主权，但1964年《赫尔辛基宣言》的建立标志着向统一伦理框架方向的国际共同努力。这一宣言经过多次修订，强调了开始任何研究项目之前进行科学审查的必要性。

1979年《贝尔蒙特报告》的发布是另一个分水岭时刻，它凸显了个别研究者在承诺伦理标准上的不足。这促使全球政府介入，颁布和执行旨在促进负责任研究行为的规定。

2010年，《新加坡研究诚信声明》在全球代表的集会中产生，旨在寻求一种和谐的调控方式，尊重各国独立的法律。在这一背景下定义的研究诚信，超越了仅仅遵守外部命令的层面，强调了贯穿研究生命周期——从规划到传播——的伦理原则和专业标准的内在价值。

负责任研究的格局受到一系列指导方针的影响，包括专业守则、政府规章、机构政策，以及至关重要的研究者自身的个人价值观和伦理信念。这些指南的核心是确保研究负责任进行的基本伦理原则。

（1）诚实：在研究的所有方面，从数据报告到与同行及公众的交流，都要保持真实。

（2）客观性：在所有研究阶段减少偏见，承认可能影响结果的个人或财务利益。

（3）正直：坚守承诺，真诚行事，确保思想与行动的一致性。

（4）谨慎：勤勉以避免错误，并彻底审查个人及同事的工作。

（5）开放性：培养透明和接受批评的文化，公开分享资源和发现。

（6）知识产权：尊重版权和其他知识产权，适当地归功于所有贡献。

（7）保密性：保护敏感信息，尊重所有利益相关者的隐私和信任。

（8）负责任的发布和指导：出版旨在推进知识，以诚信进行指导，促进学生和同行的发展。

（9）尊重、社会责任和非歧视：尊重同事，致力于社会福利，确保所有交易的公平和平等。

（10）能力和合法性：不断寻求个人和专业成长，同时遵守法律和监管标准。

（11）对动物和人类受试者的关怀：以最高的尊重和伦理考虑进行涉及动物和人类的研究，优先考虑他们的福祉和权利。

研究伦理和诚信的这种进化反映了对一种平衡科学探究与社会及自然世界福祉需要的原则化方法日益增长的认识。

Chapter 4　　Chemistry Reactions

4.1　　Displacement reactions

What is a displacement reaction?

A displacement reaction, often characterized in chemistry, involves the substitution of an atom or group of atoms by another atom within a compound. This type of reaction is illustrated by the interaction between iron and a copper sulphate solution, where iron displaces the copper.

Example of displacement reaction:
$$A + B - C \longrightarrow A - C + B$$

In the reaction formula above, "A" represents a more reactive element than "B". For this reaction to occur, A and B must be either halogens (where "C" is a cation) or different metals (where "C" is an anion).

Single displacement reaction

Also known as a single replacement reaction, this is a specific type of oxidation-reduction reaction. It involves an element or ion moving out of a compound, resulting in the substitution of one element for another within the compound. This substitution is dependent on the reactivity of the elements involved; a more reactive element will replace a less reactive one.

Example: When gaseous chlorine is introduced to a solution of sodium bromide, it displaces the bromine due to its higher reactivity. The solution changes color indicating the displacement:

chlorine + sodium bromide ⟶ sodium chloride + bromine

$$Cl_2(aq) + 2NaBr(aq) \longrightarrow 2NaCl(aq) + Br_2(aq)$$

Another instance involves dissolving silver nitrate in water and introducing a copper wire. Over time, silver deposits on the copper, and the solution turns bluish as copper ions develop:

$$Cu(s) + 2AgNO_3(aq) \longrightarrow 2Ag(s) + Cu(NO_3)_2(aq)$$

Double displacement reaction

This reaction occurs when components of two ionic compounds exchange parts, forming two new compounds. It typically takes place in aqueous solutions, leading to precipitation and ionic exchange.

Example: The reaction between lead (Ⅱ) nitrate and potassium iodide in an aqueous solution results in the formation of potassium nitrate (soluble) and lead iodide (insoluble), the latter precipitating out of the solution:

$$Pb(NO_3)_2(aq) + 2KI(aq) \longrightarrow 2KNO_3(aq) + PbI_2(s)$$

Another common reaction occurs when barium chloride mixes with sodium sulphate, rapidly forming a white precipitate of barium sulphate. This transformation is a hallmark of ionic reactions, characterized by the dissolution of reactants into ions and their subsequent exchange in the solution:

$$BaCl_2(aq) + Na_2SO_4(aq) \longrightarrow 2NaCl(aq) + BaSO_4(s)$$

Through these examples, the principles of displacement and double displacement reactions highlight fundamental aspects of chemical reactivity and ionic exchange in solutions.

New words and expressions

displacement reaction　置换反应
copper sulphate solution　硫酸铜溶液
single displacement reaction　单置换反应
single replacement reaction　单替换反应
oxidation-reduction reaction　氧化还原反应
element　元素
reactive　反应性
sodium bromide　溴化钠
sodium chloride　氯化钠
silver nitrate　硝酸银
copper ions　铜离子
double displacement reaction　双置换反应

ionic compounds　离子化合物
aqueous solutions　水溶液
precipitation　沉淀
ionic exchange　离子交换
lead（Ⅱ）nitrate　硝酸铅（Ⅱ）
potassium iodide　碘化钾
potassium nitrate　硝酸钾
lead iodide　碘化铅
barium chloride　氯化钡
sodium sulphate　硫酸钠
barium sulphate　硫酸钡

译文

什么是置换反应？

置换反应是化学中的一种反应，涉及一个原子或一组原子在化合物中被另一个原子替换。这种反应的一个例子是铁与硫酸铜溶液的相互作用，其中铁置换了铜。

置换反应的例子：

$$A + B - C \longrightarrow A - C + B$$

式中，"A"表示比"B"更活泼的元素。此反应发生的条件是，A 和 B 必须是卤素（其中"C"是阳离子）或不同的金属（其中"C"是阴离子）。

单置换反应

单置换反应也被称为单替换反应，这是一种特定类型的氧化还原反应。它涉及一个元素或离子从化合物中移出，导致化合物中一个元素被另一个元素替换。这种替换取决于涉及元素的反应性，较活泼的元素将替换较不活泼的元素。

例子：当气态氯被引入溴化钠溶液中时，由于氯更活泼，它会置换溴。溶液颜色的变化表明了置换发生：

$$氯 + 溴化钠 \longrightarrow 氯化钠 + 溴$$

$$Cl_2(aq) + 2NaBr(aq) \longrightarrow 2NaCl(aq) + Br_2(aq)$$

另一个例子是将硝酸银溶解在水中，并将铜丝浸入其中。随着时间的推移，铜丝上会沉积银，并且随着铜离子的形成，溶液变成蓝色：

$$Cu(s) + 2AgNO_3(aq) \longrightarrow 2Ag(s) + Cu(NO_3)_2(aq)$$

双置换反应

这种反应发生在两个离子化合物的成分互换部分,形成两个新的化合物。它通常发生在水溶液中,导致沉淀和离子交换。

例子:在水溶液中,硝酸铅(Ⅱ)和碘化钾的反应生成可溶的硝酸钾和不溶的碘化铅,后者从溶液中沉淀出来:

$$Pb(NO_3)_2(aq) + 2KI(aq) \longrightarrow 2KNO_3(aq) + PbI_2(s)$$

另一个常见的反应是氯化钡与硫酸钠混合,迅速形成白色硫酸钡沉淀。这种转变是离子反应的特征,其特点是反应物溶解成离子,并在溶液中进行离子交换:

$$BaCl_2(aq) + Na_2SO_4(aq) \longrightarrow 2NaCl(aq) + BaSO_4(s)$$

通过这些例子,置换和双置换反应的原理凸显了化学反应性和溶液中的离子交换的基本方面。

4.2 Redox reaction

A redox reaction is an important type of chemical reaction. In such a reaction, one reactant is oxidized and another reactant is reduced.

Oxidation was once regarded as a chemical reaction in which oxygen was added to a substance, whereas reduction occurred when oxygen was lost. Oxidation occurs when electrons are lost, reduction occurs when electrons are gained. To help you remember: Oil-Rig (Oxidation is electron loss, reduction is electron gain).

The oxidation number or oxidation state of an atom is a positive or negative number which is decided using agreed rules: The oxidation number of an atom of a free element is zero; The oxidation number of the ion of an element is equal to its charge; The algebraic sum of the oxidation numbers of the atoms in the formula of an electrically neutral compound is zero. For $Ba(OH)_2$, for example, the sum of the oxidation numbers is zero; For convenience, shared electrons in covalent compounds are assigned to the element having the greater electronegativity. For example, the oxidation number for phosphorus in PCl_3 is +3 (chlorine is +1), since phosphorus is less electronegative than chlorine; the algebraic sum of the oxidation numbers of all the atoms of an ion is equal to the charge on the ion. In SO_3^{2-}, for example, the sum of the oxidation numbers is [+6 for S] + [3(-2) for O] = +2. Remember that: During chemical reactions an increase in oxidation number signifies OXIDATION; a decrease REDUCTION.

Oxidation numbers are used to widen the definition of oxidation and reduction, with oxidation being defined as an increase in oxidation number and reduction as a decrease in oxidation number. Note that the oxidation number system is just designed to help you work out whether substances are oxidized or reduced. It does not tell you anything about the bonding in a compound. For example, the fact that carbon has an oxidation number of during chemical reactions an increase in oxidation number signifies +4 in some of its compounds does not mean that it exists as +4 ions in these compounds.

An oxidizing agent is a substance that takes up electrons during a chemical reaction and, in doing so, becomes reduced. A reducing agent supplies the electrons in this process and so becomes oxidized.

New words and expressions

redox reaction　　氧化还原反应　　　　oxidation state　　氧化状态
oxidized　　氧化　　　　　　　　　　　electronegativity　　电负性
reduced　　还原　　　　　　　　　　　covalent compounds　　共价化合物
oxidation　　氧化　　　　　　　　　　oxidizing agent　　氧化剂
reduction　　还原　　　　　　　　　　reducing agent　　还原剂
oxidation number　　氧化数

译文

氧化还原反应是一种重要的化学反应类型。在这种反应中，一个反应物被氧化，另一个反应物被还原。氧化过去被视为一种向物质中添加氧的化学反应，而失去氧则发生还原。氧化是失去电子，还原是得到电子。帮助记忆的方法：Oil-Rig（氧化是失电子，还原是得电子）。

原子的氧化数或氧化状态是一个正或负的数字，这个数字是根据约定的规则确定的：自由元素的原子的氧化数为零；元素离子的氧化数等于其电荷；在电中性化合物的公式中，原子的氧化数的代数和为零。例如，对于$Ba(OH)_2$，氧化数之和为零；为方便起见，在共价化合物中共享的电子分配给电负性较大的元素。例如，在 PCl_3 中，磷的氧化数为 +3（氯为 +1），因为磷的电负性小于氯，所有原子的氧化数的代数和等于该离子上的电荷。例如，在 SO_3^{2-} 中，氧化数之和为 [对 S 为 +6] + [对 O 为 3(-2)] = +2。记住：在化学反应中，氧化数的增加表示氧化；减少表示还原。

氧化数的使用扩展了氧化和还原的定义，氧化被定义为氧化数的增加，还原则为氧化数的减少。注意，氧化数系统只是设计来帮助你判断物质是被氧化还是被还原，它不会告诉你化合物中的键合信息。例如，碳在某些化合物中的氧化数为 +4，并不意味着它以 +4 离子的形式存在于这些化合物中。

氧化剂是在化学反应中吸收电子的物质，因而被还原。还原剂在此过程中提供电子，因而被氧化。

4.3　Types of chemical reactions

Combustion reaction

A combustion reaction occurs when a combustible substance reacts with an oxidizing agent, typically oxygen, to produce an oxidized product along with the release of energy, usually in the form of heat. A prime example is the combustion of magnesium metal：

$$2Mg + O_2 \longrightarrow 2MgO + heat$$

In this reaction, two atoms of magnesium combine with one molecule of oxygen to form two molecules of magnesium oxide, releasing energy as heat.

Decomposition reaction

Decomposition reactions involve a single substance breaking down into two or more products, often requiring an energy input such as heat, light, or electricity to break chemical bonds. An example is the decomposition of calcium carbonate, a critical ingredient in cement, into calcium oxide (quicklime) and carbon dioxide when heated：

$$CaCO_3(s) \xrightarrow{heat} CaO(s) + CO_2(g)$$

Here, calcium carbonate undergoes thermal decomposition into calcium oxide and carbon dioxide gas.

Neutralization reaction

Neutralization reactions occur between an acid and a base, producing salt and water. In these reactions, water forms from the combination of hydroxide (OH^-) ions and hydrogen (H^+) ions. The resultant pH of a reaction between a strong acid and a strong base will be neutral ($pH = 7$). For instance, the reaction between hydrochloric acid and sodium hydroxide yields sodium chloride (common salt) and water:

$$HCl + NaOH \longrightarrow NaCl + H_2O$$

This exemplifies how hydrochloric acid and sodium hydroxide react to produce common salt and water.

Precipitation or double-displacement reaction

This type of reaction is a subset of displacement reactions where ions of two compounds exchange places, forming two new products. A notable instance is the reaction between silver nitrate and sodium chloride, resulting in the formation of silver chloride and sodium nitrate:

$$AgNO_3 + NaCl \longrightarrow AgCl + NaNO_3$$

In this double displacement reaction, silver displaces sodium in sodium chloride, forming silver chloride and sodium nitrate.

Synthesis reaction

Synthesis reactions involve multiple simple substances combining under specific conditions to form a more complex product. The resulting product is always a compound. A classic example is the synthesis of sodium chloride from solid sodium and chlorine gas:

$$2Na(s) + Cl_2(g) \longrightarrow 2NaCl(s)$$

Here, solid sodium reacts with chlorine gas to produce sodium chloride, commonly known as table salt.

Important points to remember:

(1) Chemical vs. physical change: In a chemical change, a new compound is formed, whereas in a physical change, the substance merely changes its state of existence.

(2) Reactants and products: Reactants are the atoms, ions, or molecules that participate in a reaction to form new substances known as products.

(3) Law of conservation of mass: This fundamental principle of chemistry states that in a chemical reaction, no atoms are destroyed or created; they are merely rearranged to form new products from the reactants.

New words and expressions

combustible　可燃的
combustion reaction　燃烧反应
oxidizer　氧化剂
oxidized product　氧化产物
magnesium metal　镁金属
magnesium oxide　氧化镁
decomposition reaction　分解反应
calcium carbonate　碳酸钙
calcium oxide (quick lime)　氧化钙（生石灰）
carbon dioxide　二氧化碳
neutralization reaction　中和反应
hydrochloric acid　盐酸
sodium hydroxide　氢氧化钠
sodium chloride (common salt)　氯化钠（食盐）
double-displacement reaction　双置换反应
precipitation　沉淀
silver chloride　氯化银
sodium nitrate　硝酸钠
synthesis reaction　合成反应
chlorine gas　氯气
reactants　反应物
products　产物
law of conservation of mass　质量守恒定律

译文

燃烧反应

燃烧反应是一种可燃物质与氧化剂（通常是氧气）反应生成氧化产物并释放能量（通常为热能）的化学反应。例如，镁金属的燃烧反应：

$$2Mg + O_2 \longrightarrow 2MgO + 热能$$

在此反应中，两个镁原子与一个氧分子反应生成两个氧化镁分子，并在过程中释放热能。

分解反应

分解反应是单一物质分解成多个产物的过程，通常需要如热、光或电等能量的输入来打破化合物的键。一个典型的例子是加热碳酸钙（水泥的主要成分之一），生成氧化钙（生石灰）和二氧化碳：

$$CaCO_3(s) \xrightarrow{热} CaO(s) + CO_2(g)$$

在这里，加热的碳酸钙分解成氧化钙和二氧化碳气体。

中和反应

中和反应是酸和碱反应生成盐和水的过程。在这种反应中，水是通过氢离子（H^+）和氢氧根离子（OH^-）结合而形成的。当强酸和强碱进行中和反应时，最终的pH值将是中性的（pH=7）。例如，盐酸与氢氧化钠反应生成氯化钠（普通食盐）和水：

$$HCl + NaOH \longrightarrow NaCl + H_2O$$

这说明了盐酸和氢氧化钠如何反应生成食盐和水。

双置换反应

双置换反应是置换反应的一种，其中两种化合物的离子互换位置，形成两个新的产物。一个显著的例子是硝酸银与氯化钠反应，生成氯化银和硝酸钠：

$$AgNO_3 + NaCl \longrightarrow AgCl + NaNO_3$$

在这个双置换反应中，银离子置换了氯化钠中的钠离子，形成了氯化银和硝酸钠。

合成反应

合成反应是多种简单物质在特定条件下结合形成更复杂产物的过程。产生的产物总是化合物。一个经典的例子是固态钠与氯气合成氯化钠：

$$2Na(s) + Cl_2(g) \longrightarrow 2NaCl(s)$$

这里，固态钠与氯气反应生成氯化钠，即我们熟知的食盐。

需要记住的重点：

（1）化学变化与物理变化：在化学变化中，形成新的化合物；而在物理变化中，物质只是改变其存在状态。

（2）反应物与产物：参与反应形成新物质的原子、离子或分子称为反应物；新形成的原子或分子称为产物。

（3）质量守恒定律：这是化学的基本原理，指出在化学反应中，没有原子被摧毁或创造；它们只是重新排列，从反应物形成新的产物。

4.4　Electrolysis（battery）

Electrolysis involves the flow of electrical current through a liquid, leading to chemical changes. This process requires a conducting liquid, such as a molten ionic compound or an aqueous solution, capable of carrying an electrical charge.

Consider the example of sodium chloride. When heated to over 801 degrees Celsius, it melts into a liquid containing freely moving sodium ions (cations) and chloride ions (anions). Introducing copper electrodes into this molten salt and connecting them to a power source sets the stage for fascinating chemical interactions. Copper is chosen for its high melting point and good conductivity.

In this setup, electrons in the copper move freely, forming a "sea" that drifts toward the positive terminal of a battery when a voltage is applied. This electron flow constitutes the current—essentially the movement of electrical charge.

Turning our attention back to the molten sodium chloride now connected to an electrical cell, the flowing electrons induce a charge in the electrodes—one becomes positive and the other negative. This charge differential attracts the sodium ions to the negative electrode and chloride ions to the positive electrode, demonstrating the principle that opposite charges attract.

Upon contact with the electrodes, the ions undergo electron exchange, sparking chemical reactions. These reactions can produce solids, release gases, or leave behind new solutions, depending on the electrodes' material and the liquid's properties.

The electrodes, often larger metal plates, play crucial roles: The anode (positive electrode) attracts anions, and the cathode (negative electrode) attracts cations. This configuration enables the efficient conduction of electricity through the liquid, now termed an electrolyte. Effective electrolysis also depends on the concentration of the solution and the applied voltage, which influence the rate at which ions flow.

In the context of aluminum extraction—a process critical despite its high energy cost—aluminum oxide, a white powder obtained from bauxite ore, must be melted for electrolysis. Due to its high melting point (over 2000 degrees Celsius), aluminum oxide is dissolved in molten cryolite, which has a much lower melting point, reducing energy costs. In this molten mixture, graphite electrodes facilitate the extraction: Aluminum forms at the cathode and sinks due to its higher density, while oxygen forms at the anode, reacting with carbon to produce carbon dioxide and gradually depleting the graphite anode.

This brief overview touches on key aspects of electrolysis, emphasizing its role in chemical transformation and industrial applications like aluminum extraction. The movement of ions in an electrolyte contrast with the flow of electrons in electrodes and wires, highlighting electrolysis's dual nature in conducting electricity and driving chemical reactions.

New words and expressions

electrolysis　电解
electrical current　电流
ionic compound　离子化合物
aqueous solution　水溶液
sodium ions　钠离子
chloride ions　氯离子
electrodes　电极
electrical cell　电池
anode　阳极
cathode　阴极

electron exchange　电子交换
electrolyte　电解质
voltage　电压
aluminum oxide　氧化铝
bauxite ore　铝矿石
extraction　提取
cryolite　冰晶石
graphite　石墨
carbon dioxide　二氧化碳

译文

电解涉及电流通过液体流动，引起化学变化。这一过程需要一个能够传导电流的液体，如熔融的离子化合物或水溶液。

以氯化钠为例。当其加热至超过801 ℃时，它会融化成含有自由移动的钠离子（阳离子）和氯离子（阴离子）的液体。在这种熔融盐中引入铜电极，并将它们连接到电源，就为化学相互作用搭建了舞台。选择铜是因为它具有高熔点和良好的导电性。

在此设置中，铜中的电子可以自由移动，形成一个"电子海"，当施加电压时，这些电子会向电池的正极漂移。这种电子流动构成了电流——本质上是电荷的移动。

将注意力重新转向现在连接到电池的熔融氯化钠，流动的电子会使电极带电——一个变正，另一个变负。这种电荷差吸引钠离子向负电极移动，氯离子向正电极移动，展示了相反电荷相吸的原理。

当离子与电极接触时，会发生电子交换，激发化学反应。这些反应可以产生固体，释放气体或留下新的溶液，这取决于电极的材料和液体的属性。

电极，通常是较大的金属板，扮演着关键的角色：阳极（正电极）吸引阴离子，而阴极（负电极）吸引阳离子。这种配置使液体能有效地传导电流，现在被称为电解质。有效的电解还依赖于溶液的浓度和施加的电压，这些因素影响离子的流动速率。

在铝提取的背景下——尽管其能耗很高，这一过程仍然至关重要——从铝矿石中得到的白色粉末氧化铝必须熔化以进行电解。由于氧化铝的熔点非常高（超过2000 ℃），人们通常将其溶解在熔融的冰晶石中，后者的熔点远低于氧化铝，这降低了能源成本。在这种熔融混合物中，石墨电极促进了提取过程：铝在阴极形成并因其密度较高而沉淀，而氧气在阳极形成，与石墨反应生成二氧化碳，并逐渐消耗掉石墨阳极。

这一简要概述涵盖了电解过程中的关键要素，强调了其在化学转化和工业应用（如铝提取）中的重要作用。电解质中离子的运行与电极和导线中电子的流动形成了鲜明对比，突显出电解过程在导电与化学反应推动中的双重特性。

4.5　Corrosion of iron

Many metals, including iron, are susceptible to corrosion when they come into contact with air and water. The reactivity of these metals can often be determined using their reduction potentials. In the presence of oxygen and water, iron undergoes oxidation from its elemental form to Fe^{2+}, and can further oxidize to Fe^{3+}. These oxidation processes lead to the formation of rust, which is chemically identified as $Fe_2O_3 \cdot xH_2O$. The "x" in this formula denotes that rust has a variable water content. Rust formation requires three crucial components: oxygen, water, and electrolytes dissolved in the water. Without any one of these components, rusting significantly slows down or does not occur.

To protect iron or steel from corrosion, several methods can be implemented:

(1) Surface protection: Applying paint, oil, grease, or a plastic coating shields the metal surface from direct exposure to air and water, thus preventing rust.

(2) Galvanization: Coating iron with a more reactive metal like zinc, which has a more negative standard electrode potential ($E°$), ensures that zinc will corrode preferentially if the coating is damaged. This method involves either dipping the iron into molten zinc or electroplating it. The zinc layer, once exposed to air, forms a protective barrier of zinc oxide, which inhibits further corrosion. This process is termed "galvanizing" and helps the zinc to become passive, thereby extending the life of the iron or steel beneath.

(3) Sacrificial protection: For larger structures such as ships or pipelines, it is often impractical to apply galvanization uniformly. Instead, blocks of a more reactive metal like magnesium or zinc are attached to the object. These sacrificial anodes are designed to corrode preferentially, thereby protecting the primary metal from oxidation and corrosion. This technique provides a long-term defense against environmental factors that cause corrosion.

These preventative strategies are essential in extending the durability and functionality of metallic structures, particularly those exposed to harsh environmental conditions.

New words and expressions

reduction potentials   还原电位   formula   化学式
oxidize   氧化   corrosion   腐蚀
Fe^{2+} 二价铁   galvanizing   镀锌
Fe^{3+} 三价铁   sacrificial protection   牺牲保护
rust   铁锈   zinc oxide   氧化锌

reactive metal	活泼金属	electroplate	电镀
standard electrode potential (E°)	标准电极电位	durability	耐久性
		functionality	功能
sacrificial anodes	牺牲阳极		

译文

许多金属，包括铁，在与空气和水接触时容易腐蚀。这些金属的反应性通常可以通过其还原电位来确定。在氧气和水的存在下，铁从元素形式氧化为 Fe^{2+}，并可以进一步氧化为 Fe^{3+}。这些氧化过程导致了铁锈的形成，化学上标识为 $Fe_2O_3 \cdot xH_2O$。公式中的"x"表示铁锈具有可变的水含量。形成铁锈需要三个关键成分：氧气、水和溶解在水中的电解质。如果缺少这些成分中的任何一个，锈蚀就会大大减慢或不发生。

为了保护铁或钢铁免受腐蚀，可以实施几种方法：

（1）表面保护：通过涂漆、涂油、涂脂或涂塑料涂层保护金属表面，避免直接暴露在空气和水中，从而防止锈蚀。

（2）镀锌：用锌等更活泼的金属（其标准电极电位 E° 更负）覆盖铁，确保即使涂层受损，锌也会优先失去电子。这种方法包括将铁浸入熔融锌中或对铁进行电镀。锌层一旦暴露于空气中，会形成一层保护性的氧化锌屏障，阻止进一步腐蚀。这个过程被称为"镀锌"，有助于使锌被动化，从而延长其下方的铁或钢铁的使用寿命。

（3）牺牲保护：对于像船只或管道这样的大型结构，通常不切实际地均匀应用镀锌。相反，将更活泼的金属（如镁或锌）的块状材料固定在物体上。这些牺牲阳极被设计为优先腐蚀，从而保护主金属免受氧化和腐蚀。这种技术为抵御引起腐蚀的环境因素提供了长期防御。

这些预防策略对于延长金属结构的耐用性和功能性至关重要，尤其是那些暴露在恶劣环境条件下的结构。

4.6 Phenomena associated with reactions

4.6.1 Exothermic and endothermic reactions

This discussion will explore exothermic and endothermic reactions, focusing on their distinctions and the associated energy changes.

An exothermic reaction releases energy to its surroundings, exemplified by a fire emitting heat. Conversely, an endothermic reaction absorbs energy from its surroundings, as demonstrated by a melting snowman which absorbs heat. In terms of terminology, "exo" refers to external, indicating energy release, while "endo" denotes internal, indicating energy absorption.

Exothermic reactions, which transfer heat energy to the surroundings, often result in an increase in ambient temperature, akin to the warmth provided by a bonfire. Common examples include neutralization reactions between acids and alkalis, interactions between water and calcium oxide, and cellular respiration. These reactions are readily identifiable through the use of a thermometer, which will register an increase in temperature upon the mixing of hydrochloric acid and sodium hydroxide, for instance. The prevalence of exothermic reactions is due to the liberation of heat during these processes. Additionally, certain physical processes, such as freezing and condensation, are exothermic because they involve bond formation, necessitating energy expenditure.

In exothermic reactions, the initial energy of the reactants is higher than that of the products, resulting in a downward trajectory on an energy curve.

Turning to endothermic reactions, these are characterized by their absorption of heat from the surroundings, making them less common. Such reactions typically result in a cooling of the environment, which can be measured with a thermometer. Examples include electrolysis, the reaction between sodium carbonate and ethanoic acid, and photosynthesis. Physical processes such as melting and boiling also exemplify endothermic reactions, where energy must be input to break bonds, transitioning from solid to liquid or liquid to gas, respectively.

In endothermic reactions, the energy of the products exceeds that of the reactants, reflected by an upward slope on the energy curve.

Thus, the fundamental difference lies in the energy dynamics: Exothermic reactions release energy, heating the surroundings, while endothermic reactions absorb energy, cooling the surroundings. The detection of these reactions can be achieved through temperature changes observed with a thermometer. Freezing and condensation are examples of exothermic processes due to energy release associated with bond formation.

Conversely, melting and boiling are endothermic due to the energy required to break molecular bonds.

New words and expressions

exothermic reactions　放热反应
endothermic reactions　吸热反应
energy　能量
heat　热能
neutralization reactions　中和反应
cellular respiration　细胞呼吸
thermometer　温度计
hydrochloric acid　盐酸
alkali　碱

sodium hydroxide　氢氧化钠
electrolysis　电解
photosynthesis　光合作用
freezing　结冰
condensation　凝结
melting　熔化
boiling　沸腾
energy curve　能量曲线
ethanoic acid　乙酸

译文

本讨论将探讨放热反应和吸热反应，重点关注它们的区别和相关的能量变化。

放热反应向周围环境释放能量，如火焰散发热能。相反，吸热反应从周围环境吸收能量，例如正在融化的雪人吸收热能。在术语上，"exo"表示外部，意味着释放能量；而"endo"表示内部，意味着吸收能量。

放热反应将热能传递给周围环境，通常导致环境温度升高，就像篝火为周围人提供温暖一样。常见的例子包括酸碱中和反应、水和氧化钙之间的反应及细胞呼吸。这些反应可以通过温度计来轻松识别，例如在混合盐酸和氢氧化钠时，温度计将显示温度升高。大多数化学反应是放热的，因为这些过程中释放了热能。此外，某些物理过程如结冰和凝结也是放热的，因为它们涉及形成键合，需要消耗能量。

在放热反应中，反应物的初始能量高于产物，导致能量曲线向下。

转向吸热反应，这些反应的特点是从周围环境吸收热能，使它们不太常见。这类反应通常导致环境冷却，这可以通过温度计来测量。例如电解、碳酸钠和乙酸反应及光合作用。物理过程中的熔化和沸腾也是吸热反应的例子，其中能量必须输入以打破键，从固体转变为液体或从液体转变为气体。

在吸热反应中，产物的能量高于反应物，能量曲线呈上升趋势。

因此，两者之间的基本区别在于能量动态：放热反应释放能量，使周围环境变热；而吸热反应吸收能量，使周围环境变冷。这些反应可以通过温度计观察到的温度变化来检测。结冰和凝结是放热过程的例子，因为需要形成键合，释放出能量。相反，熔化和沸腾是吸热的，因为打破键需要额外的能量。

4.6.2 Crystallization

Crystallization is the process in which solids crystallize to form a highly organized structure known as a crystal. Crystals can form through several methods, such as precipitating from a solution, freezing, or, less commonly, by deposition directly from a gas phase. The characteristics of the resulting crystals are influenced by various factors including temperature, air pressure, the rate of cooling, and in the case of liquid crystals, the evaporation duration of the fluid.

The process of crystallization comprises two primary phases. The initial phase, nucleation, involves the emergence of a crystalline phase from a supercooled liquid or a supersaturated solution. Following nucleation is the crystal growth phase, where there is an increase in particle size that solidifies into a crystal structure. During this phase, particles arrange themselves in layers on the crystal's surface, fitting into any available gaps such as pores or cracks.

In the context of chemical engineering, crystallization is used as a solid-liquid separation technique, allowing for the mass transfer of a solute from a liquid solution to a pure solid crystalline phase. Typically, this process takes place within a device known as a crystallizer. Crystallization is closely related to precipitation, but unlike precipitation, which results in an amorphous or disordered structure, crystallization results in a well-ordered crystal.

Some ionic compounds are defined by the formula $AB \cdot nH_2O$, where " $\cdot nH_2O$ " signifies that n molecules of water are incorporated within each unit of AB in the crystal lattice. Despite being dry to the touch because the water is integrated within the lattice, these compounds contain water of crystallization and are considered hydrated. The water can be removed by strong heating (dehydration), leaving behind an anhydrous residue. For example, blue copper (Ⅱ) sulfate crystals ($CuSO_4 \cdot 5H_2O$) turn into white, anhydrous copper (Ⅱ) sulfate upon heating: $CuSO_4 \cdot 5H_2O(s) \longrightarrow CuSO_4(s) + 5H_2O(l)$.

Reintroducing water to anhydrous copper (Ⅱ) sulfate causes it to regain its blue color. The dehydration of copper (Ⅱ) sulfate is a sequential process: (1) initially, the pentahydrate loses water to form a trihydrate; (2) followed by the formation of a monohydrate; (3) ultimately leaving the anhydrous salt; (4) Further intense heating of the anhydrous salt leads to its decomposition into copper (Ⅱ) oxide and sulfur trioxide. Other typical examples of substances containing water of crystallization include sodium carbonate decahydrate ($Na_2CO_3 \cdot 10H_2O$), commonly known as washing soda, and magnesium sulfate heptahydrate ($MgSO_4 \cdot 7H_2O$), known as epsom salts.

New words and expressions

crystallization 结晶
crystal 晶体
precipitating 沉淀
freezing 冷冻
deposition 沉积
nucleation 成核
crystal growth 晶体生长
solid-liquid separation 固液分离
solute 溶质
liquid solution 液体溶液
crystalline phase 晶体相
crystallizer 结晶器

precipitation 降水
water of crystallization 结晶水
hydrated 水合
dehydration 脱水
anhydrous 无水
copper (Ⅱ) sulfate 硫酸铜
pentahydrate 五水的
trihydrate 冰的
sodium carbonate decahydrate 碳酸钠十水合物
magnesium sulfate heptahydrate 硫酸镁七水合物

译文

结晶是一种固体通过结晶形成高度有序的晶体结构的过程。晶体可以通过几种方法形成，例如从溶液中沉淀、冷冻，或者较少见地从气态直接沉积。形成的晶体的特性受到温度、气压、冷却速率及在液晶的情况下，流体蒸发时间等因素的影响。

结晶过程包括两个主要阶段。首先是成核阶段，即从过冷的液体或过饱和溶剂中出现晶体相。接下来是晶体生长阶段，此阶段粒子大小增加并形成晶体状态。在这一阶段中，松散的粒子在晶体表面形成层，并嵌入到开放的不一致性中，如孔隙、裂缝等。

在化学工程中，结晶是一种固液分离技术，涉及从液体溶液到纯固体晶体相的溶质质量转移。通常，这一过程在称为结晶器的设备中进行。因此，结晶与沉淀相关，尽管结果不是无定形或无序的，而是形成了晶体。

一些离子化合物的公式可以写作 $AB \cdot nH_2O$，其中"$\cdot nH_2O$"表示每个 AB 单元在晶格中有 n 个水分子。由于水分子包含在晶格中，尽管这些化合物触感干燥，但它们包含结晶水，因此被认为是水合的。通过强烈加热可以驱逐水分（脱水），剩余的无水物质称为无水物。例如，加热时，蓝色的硫酸铜晶体（$CuSO_4 \cdot 5H_2O$）变为白色的无水硫酸铜：$CuSO_4 \cdot 5H_2O(s) \longrightarrow CuSO_4(s) + 5H_2O(l)$。

如果向无水硫酸铜中添加水，它会再次变蓝。通过加热脱水硫酸铜的过程是一个逐步的过程：(1) 首先，五水合物失水变为三水合物；(2) 然后形成一水合物；(3) 最终留下无水盐；(4) 如果无水盐被强烈加热，它会分解为氧化铜（Ⅱ）和三氧化硫。其他常见含结晶水的物质包括碳酸钠十水合物（洗涤苏打）：$Na_2CO_3 \cdot 10H_2O$，硫酸镁七水合物（泻盐）：$MgSO_4 \cdot 7H_2O$。

4.6.3 Sublimation

Sublimation is a phase transition in which a substance moves directly from the solid state to the gas state, bypassing the liquid phase entirely. This phenomenon is often observed in materials where the molecules gain enough energy from heat absorption to break free from their lattice structure and enter the gas phase without first becoming liquid. The act of undergoing sublimation is termed "sublime", and the substance in its gaseous form post-sublimation is referred to as a sublimate.

A notable characteristic of sublimation is the sublimation point—the specific temperature and pressure at which a substance sublimes rapidly. Common examples of sublimation include the transformation of dry ice (solid carbon dioxide) into carbon dioxide gas at room temperature and the sublimation of solid iodine into vapor when heated.

Conversely, the reverse process of sublimation is known as deposition or desublimation, where a gas turns directly into a solid without passing through a liquid phase. This transition is significant in processes like frost formation and the manufacturing of snow in artificial environments.

While all solids have the potential to sublime, most do so at such negligible rates that the process is almost imperceptible under normal conditions. Typically, a substance in the solid-state transitions through a liquid phase before becoming a gas; however, if a solid's vapor pressure exceeds the surrounding partial pressure, it can directly become a gas. This behavior is more pronounced under conditions where the temperature is just below the melting point, such as with water ice near 0 ℃.

Substances like carbon and arsenic more readily sublime from their solid states rather than evaporate from their liquid states due to the high pressures required at their triple points—the lowest pressure at which they can exist as liquids—making it challenging to obtain these substances in liquid form.

Sublimation is an endothermic process, necessitating the absorption of heat. This energy intake enables some molecules to overcome intermolecular forces and escape into the vapor phase. The enthalpy of sublimation, or the heat required for sublimation, is calculated by summing the enthalpy of fusion (solid to liquid transition) and the enthalpy of vaporization (liquid to gas transition). This measure provides a comprehensive view of the energy dynamics involved in the process of sublimation.

New words and expressions

sublimation 升华
solid state 固态
gas state 气态
liquid phase 液态
sublime 升华（动词）
sublimate 升华产物
sublimation point 升华点

desublimation 逆升华
deposition 凝华
vapor pressure 蒸汽压
triple point 三相点
enthalpy of sublimation 升华焓
enthalpy of fusion 熔化焓
enthalpy of vaporization 汽化焓

译文

升华是一种物质直接从固态转变为气态的相变过程，完全绕过液态。这种现象常见于分子从热量吸收中获得足够能量，打破其晶格结构并直接进入气态而不先变成液体的材料。经历升华的动作称为"升华"，升华后的气态物质称为升华产物。

升华的一个显著特征是升华点——物质迅速升华的特定温度和压力。常见的升华例子包括干冰（固态二氧化碳）在室温下转变为二氧化碳气体，以及加热时固态碘升华成蒸气。

相反，升华的逆过程称为凝华或逆升华，其中气体直接转变为固体，未经过液态。这一转变在霜冻形成和人造环境中制造雪的过程中很重要。

虽然所有固体都有升华的潜力，但在正常条件下，大多数固体的升华速率极低，几乎察觉不到。通常，固体物质在变成气体前会先通过液态；然而，如果固体的蒸气压超过周围同种物质的部分压力，它就可以直接变成气体。这种行为在接近熔点的温度条件下更为明显，例如接近 0 ℃ 的水冰。

如碳和砷等物质从其固态升华比从液态蒸发更容易，因为它们在相图中的三相点（物质可以以液态存在的最低压力）所需的压力非常高，使得这些物质很难以液态获得。

升华是一个吸热过程，需要吸收热量。这种能量吸收使得一些分子克服分子间力并逃逸到气相中。升华焓或所需的升华热量通过将熔化焓（固态到液态的转变）和汽化焓（液态到气态的转变）相加来计算。这一度量为升华过程中的能量动态提供了全面的视角。

4.6.4 Catalysis

Catalysis is the enhancement of a chemical reaction rate through the introduction of a substance known as a catalyst. Catalysts facilitate reactions without being consumed or altered in the process. They work by providing an alternative reaction mechanism that is often faster than the non-catalyzed route, effectively increasing the overall reaction rate without altering the potential for the original mechanism.

In catalysis, a small amount of catalyst can be highly effective if the reaction is rapid and the catalyst is quickly regenerated. Key factors that influence the rate of reaction include mixing, surface area, and temperature. Catalysts often interact with one or more reactants to form intermediate compounds, which then produce the final product while simultaneously regenerating the catalyst.

Catalysis is categorized based on the phase of the catalyst relative to the reactants. Homogeneous catalysis occurs when the catalyst and the reactants are in the same phase—typically liquid or gas—facilitating uniform interactions. In contrast, heterogeneous catalysis involves catalysts and reactants in different phases, usually with solid catalysts acting upon liquid or gaseous reactants. This category includes important catalysts like zeolites, alumina, and various metal oxides, which provide active sites for the reaction on their surfaces.

Enzymes and other biocatalysts represent a distinct category, often functioning as homogeneous catalysts if they are soluble, or as heterogeneous if they are bound to membranes. Enzymatic catalysis is pivotal in biological processes such as metabolism and catabolism, with enzymes catalyzing complex reactions under mild conditions.

In industrial applications, catalysis is indispensable across all chemical sectors, with an estimated 90% of commercially produced chemical products involving catalysts at some stage of their manufacture. These processes often leverage the unique properties of catalysts to enhance efficiency and selectivity, such as in the production of high-fructose corn syrup or acrylamide through biocatalysis.

Innovative fields like electrochemistry utilize metal-containing catalysts in fuel cells to accelerate half-reactions essential for energy conversion. Additionally, photocatalysis involves catalysts that harness light to drive redox reactions, with applications extending to dye-sensitized solar cells.

While transition metals are commonly used in catalysis, small organic molecules can also exhibit catalytic activity. These organic catalysts, including organocatalysts, operate via non-covalent interactions like hydrogen bonding, mimicking the action of metal-free enzymes. Organocatalysis distinguishes between covalent and non-covalent methods based on the nature of the catalyst-substrate interaction.

Overall, the field of catalysis continues to expand, exploring new catalysts and reaction mechanisms that offer greater efficiency, control, and environmental benefits, thereby broadening the scope of their application in both industrial and biological contexts.

New words and expressions

catalysis 催化	intermediate compounds 中间体
catalyst 催化剂	homogeneous catalysis 均相催化
reaction mechanism 反应机理	heterogeneous catalysis 异相催化

zeolites 沸石
alumina 氧化铝
biocatalysts 生物催化剂
enzymatic catalysis 酶催化
metabolism 新陈代谢
catabolism 分解代谢
electrochemistry 电化学
fuel cells 燃料电池
photocatalysis 光催化
dye-sensitized solar cells 染料敏化太阳能电池
organocatalysts 有机催化剂
non-covalent interactions 非共价相互作用
covalent 共价
hydrogen bonding 氢键
organocatalysis 有机催化
leverage 杠杆效力

译文

催化是通过引入称为催化剂的物质来增加化学反应速率的过程。催化剂促进反应而不被消耗或改变。它们通过提供一种通常比非催化途径更快的替代反应机理来工作，有效地增加了整体反应速率，而不改变原始机理的潜力。

在催化作用中，如果反应迅速且催化剂迅速再生，通常很少量的催化剂就足够有效。影响反应速率的关键因素包括混合、表面积和温度。催化剂通常与一个或多个反应物反应形成中间体，然后产生最终产品的同时再生催化剂。

催化按催化剂与反应物相对相位进行分类。均相催化发生在催化剂和反应物处于同一相态时——通常是液态或气态——促进均匀相互作用。相反，异相催化涉及催化剂和反应物处于不同相态，通常是固体催化剂作用于液体或气体反应物。这一类别包括重要的催化剂，如沸石、氧化铝和各种金属氧化物，它们在其表面提供反应的活性位点。

酶和其他生物催化剂代表了一个独特的类别，如果它们是可溶的，通常作为均相催化剂，或者如果它们与膜结合，则作为异相催化剂。酶催化在生物过程中至关重要，如新陈代谢和分解代谢，酶在温和条件下催化复杂反应。

在工业应用中，催化在所有化工行业中不可或缺，估计有90%的商业生产化学产品在其制造过程的某个阶段涉及催化剂。这些过程常利用催化剂的独特属性来提高效率和选择性，例如通过生物催化生产高果糖玉米糖浆或丙烯酰胺。

电化学的创新领域在燃料电池中使用含金属催化剂来加速对能量转换至关重要的半反应。此外，光催化涉及催化剂利用光来驱动氧化还原反应，应用范围扩展到染料敏化太阳能电池。

尽管过渡金属通常用于催化，但有机小分子也可以表现出催化活性。这些有机催化剂，包括有机催化，模仿无金属酶的作用，通过氢键等非共价相互作用。根据催化剂-底物相互作用的性质，有机催化分为共价相互作用和非共价相互作用。

总之，催化领域不断扩大，探索新的催化剂和反应机制，以提供更高的效率、控制力和环境效益，从而扩大其在工业和生物领域的应用范围。

Reference

[1] LEWIS, ROB, WYNNE EVANS. Chemistry [M]. 3rd ed. New York: Palgrave Macmillan, 2006.

[2] HOUSECROFT, CATHERINE E, ALAN G SHARPE. Inorganic Chemistry [M]. 5th ed. Harlow: Pearson Education Limited, 2018.

[3] MIESSLER, GARY L, PAUL J FISCHER, et al. Inorganic Chemistry [M]. 5th ed. Upper Saddle River: Pearson, 2013.

[4] ATKINS, PETER, TINA OVERTON, et al. Shriver and Atkins' Inorganic Chemistry [M]. 5th ed. Oxford: Oxford University Press, 2010.

[5] KITTEL, CHARLES. Introduction to Solid State Physics [M]. 8th ed. Hoboken: John Wiley & Sons, 2005.

[6] WEST, ANTHONY R. Solid State Chemistry and Its Applications [M]. 2nd ed. Chichester: Wiley, 2014.

[7] HARRIS, DANIEL C. Quantitative Chemical Analysis [M]. 9th ed. New York: W. H. Freeman and Company, 2015.

[8] SKOOG, DOUGLAS A, DONALD M WEST, et al. Fundamentals of Analytical Chemistry [M]. 9th ed. Belmont: Brooks/Cole, Cengage Learning, 2014.

[9] CHRISTIAN, GARY D, PURNENDU K DASGUPTA, et al. Analytical Chemistry [M]. 7th ed. Hoboken: John Wiley & Sons, 2014.